LABORATORY ANATOMY OF THE

TURTLE

LAURENCE M. ASHLEY, Ph.D.
Professor of Zoology
Walla Walla College
College Place, Washington

FIRST EDITION

Illustrated by CARL PETTERSON

wcb
WM. C. BROWN COMPANY PUBLISHERS
Dubuque, Iowa

Booth Laboratory Anatomy Series

Consulting Editor
Ernest S. Booth
Robert B. Chiasson

ISBN 0–697–04601–X

Fourteenth Printing, 1982

Laboratory Anatomy of the Cat
Ernest S. Booth, 1944, 1946, 1948
Robert B. Chiasson 1967, 1972

Laboratory Anatomy of the Domestic Chicken
Michael C. Robinson, 1970

Laboratory Anatomy of the Elementary Chordates
Robert B. Chiasson, 1969

Laboratory Anatomy of the Frog
Raymond A. Underhill, 1951, 1969, 1975

Laboratory Anatomy of the Human Body
Bernard B. Butterworth, 1975

Laboratory Anatomy of the Iguana
Jonathan C. Oldham, Hobart M. Smith, 1975

Laboratory Anatomy of the Mink
David Klingener, 1972

Laboratory Anatomy of Necturus
Robert B. Chiasson, 1960, 1969

Laboratory Anatomy of the Perch
Robert B. Chiasson, 1966, 1974

Laboratory Anatomy of the Fetal Pig
Theron O. Odlaug 1951, 1955, 1965, 1969

Laboratory Anatomy of the Pigeon
Robert B. Chiasson, 1959, 1972

Laboratory Anatomy of the Rabbit
Charles A. McLaughlin, 1970

Laboratory Anatomy of the White Rat
Robert B. Chiasson, 1958, 1969, 1975

Laboratory Anatomy of the Shark
Laurence M. Ashley, 1950, 1969

Laboratory Anatomy of the Turtle
Laurence M. Ashley, 1955

Printed in the United States of America

Preface

One of the most interesting vertebrates to study in comparative anatomy is the turtle. Some may think it either too difficult for class dissection or too atypical a reptile to be included in the course. Actually it requires but little dissection — mainly removal of the ventral armor (plastron) and breaking away of sufficient of the dorsal armor (carapace) to enable visualization of internal parts. Furthermore, the turtle is one of the most readily available and inexpensive of North American reptiles. Except for its armor is it an excellent example of a vertebrate whose anatomy is intermediate between that of a typical amphibian and that of a typical bird.

This manual is intended as a laboratory guide to the dissection of the turtle. Numerous labeled drawings are included so that the student may:

(1) progress rapidly and intelligently;
(2) make fewer errors in dissection;
(3) have ready access to correctly labeled drawings for review; and
(4) save valuable time which will be spent reviewing for oral quizzes covering his dissections and in making additional dissections.

It is suggested that laboratory grades be based upon quality of dissection as well as performance on oral quizzes. Students will generally appreciate not being required to make their own drawings, and grading will be more equitable when those having artistic ability are given no better chance for high grades because of making superior drawings. The time required of laboratory instructors may be reduced since oral quizzes can be given during the laboratory period and because students will have fewer questions owing to the labeled drawings.

This method of laboratory study encourages more careful dissection and observation and requires further use of the specimen dissected each time the student appears for an oral quiz. It is further suggested that instructors require students to demonstrate structures called for, rather than pointing them out for the student to name.

Constructive criticism and suggestions for the improvement of this manual will be gratefully received by the author.

L. M. Ashley.

College Place, 1954.

Introduction

Painted turtles or terrapins are commonly used in the study of the anatomy of the turtle. Turtles are reptiles belonging to Phylum CHORDATA, subphyllum Vertebrata, class Reptilia, orders Chelonia or Testudinata, families Chelydridae or Testudinidae, genera Chelydra, Chrysemys, or Pseudemys and to any one of several different species. Since a variety of species and genera are sold for dissection it is not always possible to obtain the scientific name of the turtles purchased without the use of a key and even then it must first be known just where the specimen was collected in order to know what key to use.

Before beginning to dissect you should become familiar with the following suggestions regarding laboratory work:

1. When possible, separate organs without cutting (blunt dissection).

2. Use teasing needles, probes, etc., in preference to scalpels or scissors, whenever possible.

3. Ask questions only when they cannot be answered by reading your manual.

4. When you really need help do not hesitate to call upon the instructor for assistance.

5. Variation is the rule in nature, therefore, you may expect to find differences in the anatomy of certain muscles, blood vessels, etc., from time to time. Marked deviations from the normal should be called to the attention of the instructor so that they may be demonstrated before the entire class.

The drawings in the manual represent the usual condition to be found in terrapins and in painted turtles but in most respects the anatomy of these forms is the same as that of all turtles and even tortoises, with minor exceptions.

6. You will profit by looking about at some of the dissections of other students from time to time but remember to do your own dissection so well that you can satisfactorily demonstrate any structure called for in the oral quizzes.

7. In addition to the usual dissecting tools it is desirable to have a cheap pair of pliers or a pair of bone shears for breaking off pieces of carapace when dissecting.

CHAPTER I

External Anatomy and Exoskeleton

Study the external form and structure of your turtle. Its longitudinal axis is relatively shorter than that of most vertebrates but its transverse axis is much greater. The body is so constructed as to be well adapted to life either on land or in the water. Although slow-moving on land it can swim rapidly and, its vital organs are well protected by dorsal and ventral shields, the *carapace* and the *plastron.*

A. HEAD.

Observe its reptilian form, the strong bony skull and jaw giving ample protection to the special sense organs. Note the covering of the head—is it naked, scaled, or with well-marked head shields? Is the jaw and throat region similarly covered? On the tip of the snout find the *anterior nares,* a pair of openings leading into the air passages of the respiratory tract. Observe the *horny beaks* covering the jaws. Are each of these pointed anteriorly in the midline? The horn is epidermal in origin and forms part of the exoskeleton. Examine the *eyelids,* both upper and lower, and find the *nictitating membrane* in the anterior corner of each eye. Next find the *tympanic membrane* (eardrum) on either side just behind the *angle of the jaw.* If this is not externally evident wait until the anatomy of the anterior part of the mouth is described (see Chapter IV) when it can be located by probing from within outwards.

B. NECK.

Turtles have long necks but they are especially long in snapping and in soft-shelled turtles. Note the skin covering the neck of your turtle and determine whether it is smooth or wrinkled and whether it is loose or tight.

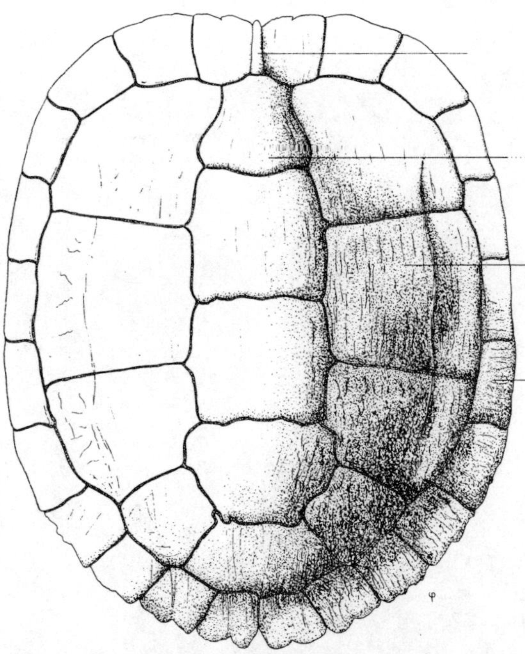

Fig. 1—Carapace, dorsal view

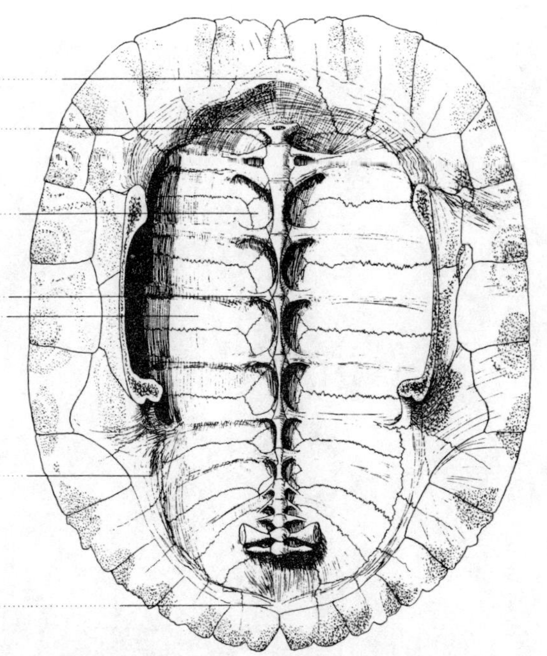

Fig. 2—Carapace, ventral view

C. TRUNK.

The trunk contains the main visceral mass of the body. It is made firm and rigid by the large bony *carapace* above, and the almost flat *plastron* below. Carapace and plastron are strongly united on each side by bony *bridges* varying in width with the species. *Horny shields* (scutes) cover these rigid structures externally forming the major portion of the exoskeleton. Learn the name and location of each kind of scute, and bony-plate (endoskeleton) of the carapace and of the plastron. Figs. 1-4.

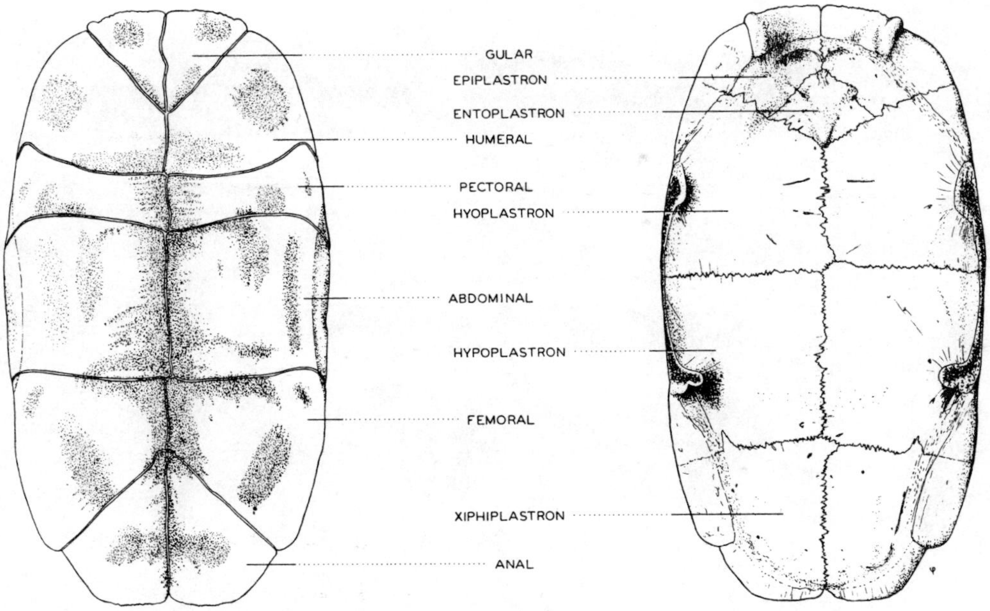

Fg. 3—Plastron, ventral view

Fig. 4—Plastron, dorsal view

D. LIMBS.

Forelimbs and hindlimbs are alike supplied with five toes each. The *toes* (digits) each exhibit a *claw* except the fifth (little) toes on the hindfeet. The legs are supplied with powerful muscles (Chapter III) in order to provide for rapid swimming, or walking on land. Spread apart the toes of a fore and hind-foot noting the webbing of skin between the toes.

E. TAIL.

This appendage is short in the painted turtle (Chrysemys) but is rather long in the snapper (Chelydra). The rounded anal opening appears on the ventral side near the base of the tail.

F. SUMMARY.

The body surface is covered with either a tough dark skin or with horny plates or shields. Horny plates cover jaws, carapace, plastron and bridges – skin covers the remaining surfaces. External orifices of *mouth*, *nostrils* (nares), and *anus* provide necessary openings for the digestive and the respiratory tracts.

CHAPTER II

Skeletal Systems (The Endoskeleton)

The first part of the skeleton, the *exoskeleton,* was considered in Chapter I. The endoskeleton is the second part of the skeletal system. It is derived partly from cartilage by replacement of cartilage by bone, and partly from membranes or connective tissues, by intramembranous ossification. For the study of the endoskeleton it is necessary to have available specially prepared skeletons. Both articulated and disarticulated skeletons are desirable. First study the articulated skeleton after which study separate bones, if these are available.

The study of the endoskeleton requires a knowledge of various bony fusions or joints. Skull bones are firmly united by immovable *suture joints* (synarthroses). Such joints consist of more or less interlocked toothed edges of the bones thus united. Joints may be slightly movable (amphiarthroses), or freely movable (diarthroses) as in the limbs.

Certain smaller or larger holes (foramina) through bones will be noticed. Only the major foramina will be mentioned in this manual. Other features of bones include such projections as spines, condyles, tuberosites, and other so-called processes of bones. These are specifically mentioned only when they seem to be of special importance in comparative anatomy.

The endoskeleton consists of two divisions, the *axial skeleton* consisting of the skull, vertebrae, and ribs, and the appendicular skeleton consisting of the limb bones together with the girdle

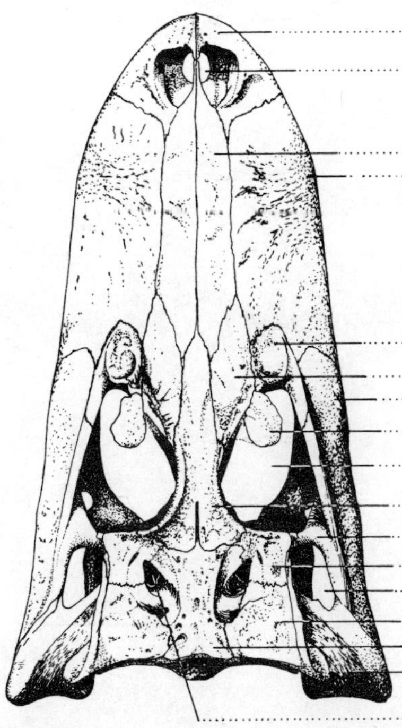

Fig. 5—Alligator skull, dorsal

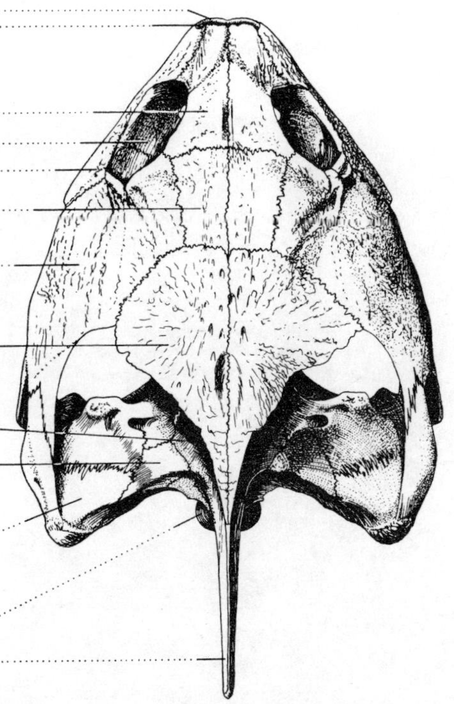

Fig. 6—Turtle skull, dorsal

bones. The girdles serve to attach the limb bones to the body or to the axial skeleton.

The bony plates of the carapace and plastron have been noted in Chapter I.

AXIAL SKELETON

A. SKULL. This consists of two portions, the *cranium* (brain case) and the *visceral skeleton,* that is, the bones and cartilages of the upper jaw, the *suspensorium* of the lower jaw, the *lower jaw* itself, and the *hyoid apparatus.*

1. CRANIUM. Study the dorsal and lateral aspects of the skull, figs. 5-8, and learn the names of the bones shown by identifying them on your turtle skull. Next study the ventral skull bones, figs. 9 and 10. The alligator skull is also shown for comparison with the turtle skull.

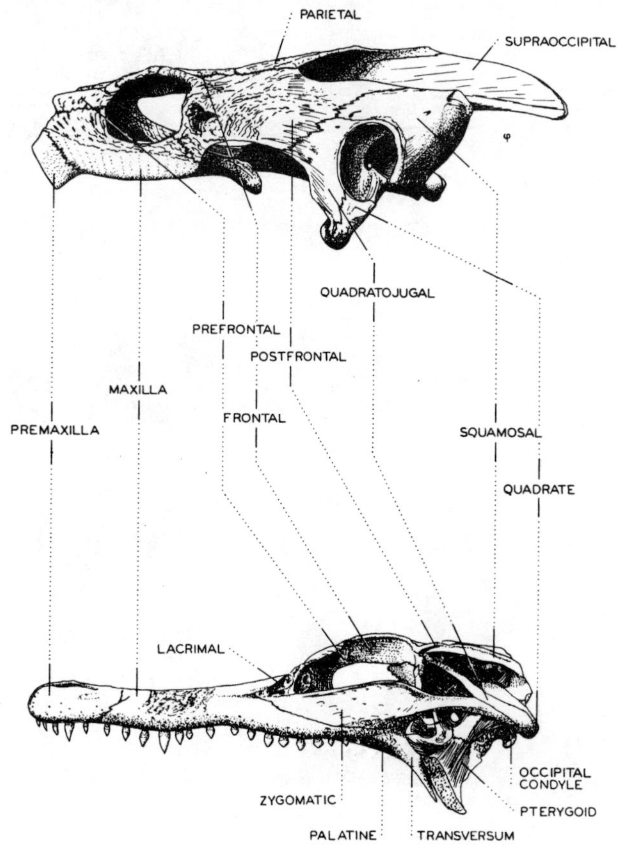

Fig. 7 and 8—Turtle and alligator skull, lateral

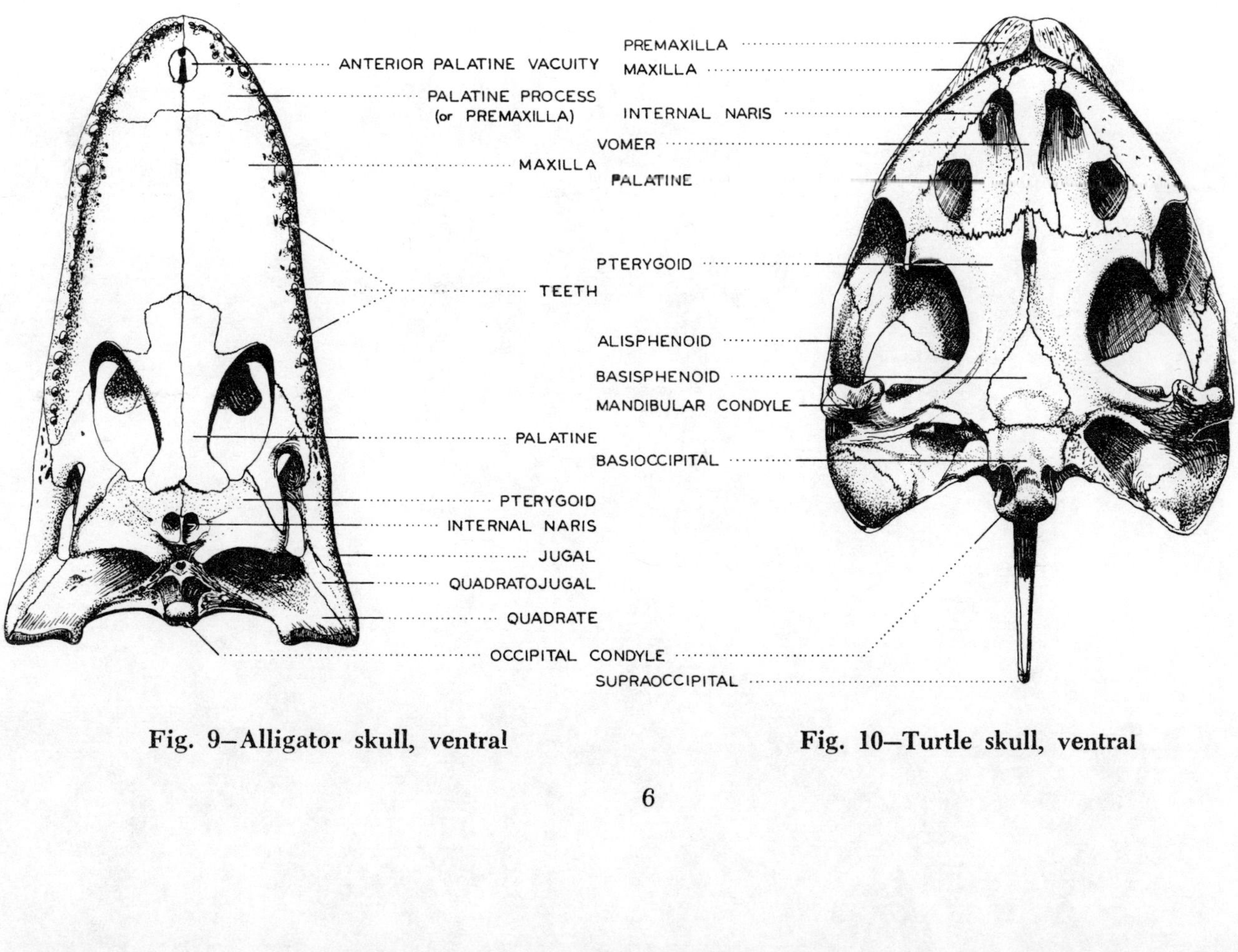

Fig. 9—Alligator skull, ventral

Fig. 10—Turtle skull, ventral

2. LOWER JAW. Study figs. 11 and 12 and identify the named bones on your skeleton. A bright light and a good hand lens may be needed to reveal some of the suture joints between the bones of the skull and of the jaw. Skulls from young turtles should reveal suture joints better than those from old turtles.

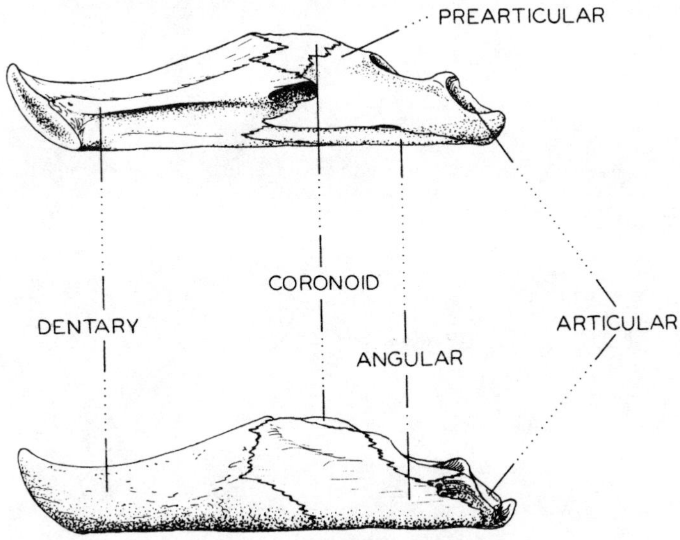

Fig. 11—Turtle jaw (above, medial; below, lateral).

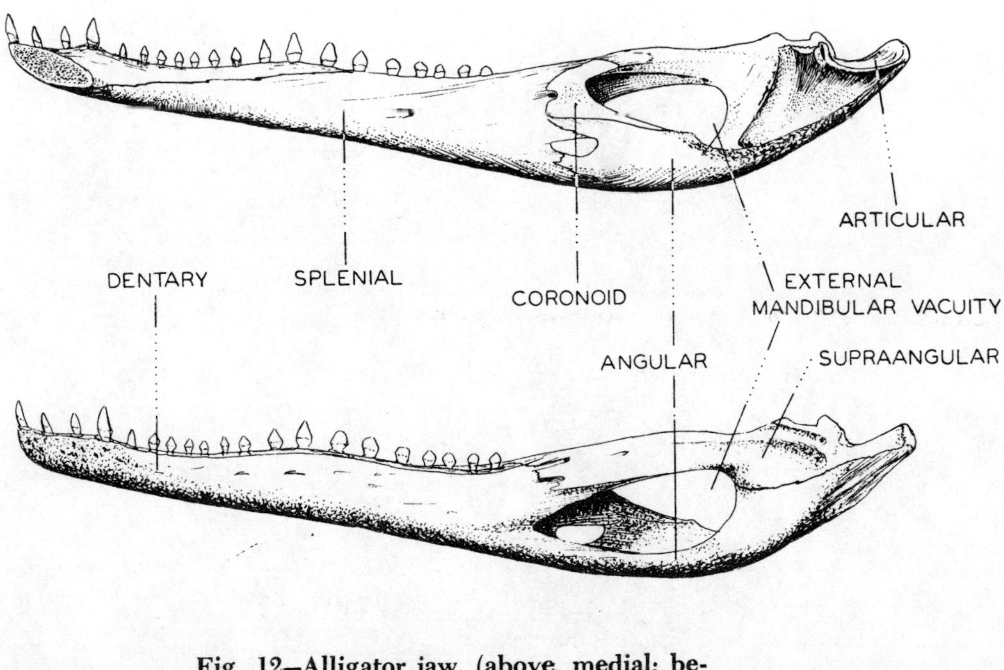

Fig. 12—Alligator jaw. (above, medial; below, lateral).

3. HYOID BONE. Normally found within the muscles of the tongue and neck, this bone derived from the second pharyngeal arch consists of an unpaired median *body* and two pairs of postero-laterally directed horns, the anterior and the posterior horns, fig. 13.

B. VERTEBRAL COLUMN AND RIBS, fig. 13.

1. CERVICAL VERTEBRAE. Eight vertebrae are present in the neck. These can be withdrawn beneath the carapace by the contraction of powerful neck muscles to be described in Chapter III. Observe the articulating processes (zygapophyses) of these large vertebrae. Notice the first cervical vertebra or *atlas*. It has little or no body or centrum because this has been added to the second vertebra, the *axis*, and forms the *odontoid process* of the latter. The odontoid process serves as a pivot about which the atlas may turn from side to side. The remaining six cervical

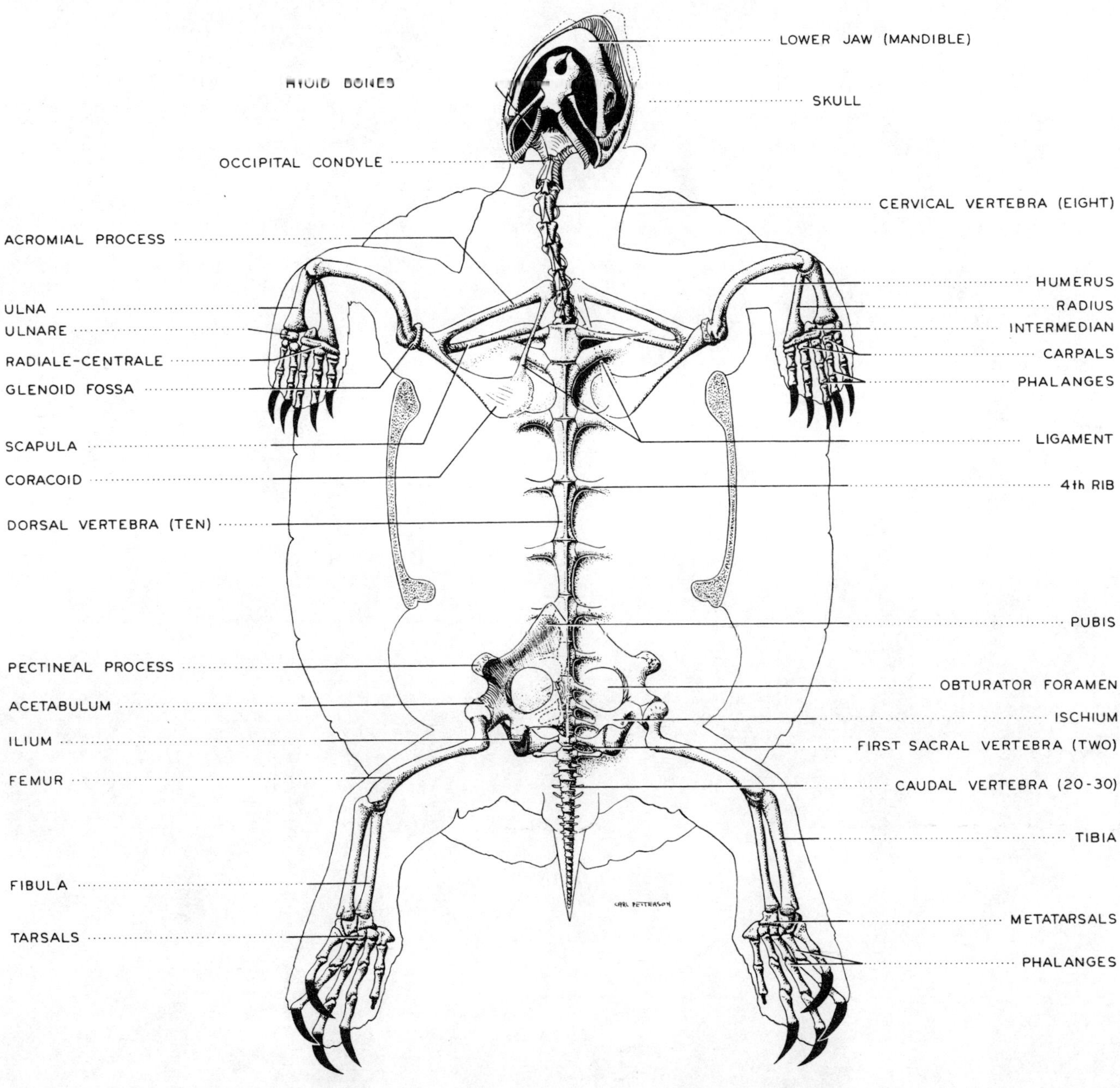

Fig. 13—Axial and appendicular skeletons, ventral

vertebrae are similar in structure and are typical of vertebrae in general.

2. DORSAL OR THORACIC VERTEBRAE. Ten of these lie caudal to the cervicals. The dorsal vertebrae are firmly fused dorsally with the vertebral plates of the carapace. These vertebrae consist of slender centra whose processes, except in the first, have been poorly developed. The first dorsal vertebra exhibits strong lateral processes curving abruptly ventrad to articulate with the eighth cervical vertebra.

3. RIBS. Each dorsal vertebra gives rise to a rib on each side. The ribs develop together with the costal plates of the carapace with which they fuse as thin bony ribbons which may reach laterad as far as the marginal plates. Anteriorly the ribs arise at right angles from the intervertebral joints but posteriorly they arise from stubby lateral bases of the vertebrae and project posterolaterally from the centra. Some variation from this pattern may be seen in different species of turtles. Notice that the first pair of ribs, being shorter and smaller, fuses with the second pair, a short distance from vertebrae one and two of the dorsal group.

4. SACRAL VERTEBRAE. Dorsal vertebrae eleven and twelve are the sacral vertebrae and are not fused with the carapace. The first exhibits enlarged ribs, somewhat pyramidal in shape, whose expanded distal ends are fused to the ilia of the pelvic girdle.

5. CAUDAL VERTEBRAE. From twenty-five to thirty caudal vertebrae generally are present. Like the cervicals, the caudals are free from the carapace and are moved, in life, by a number of slender tail muscles. Caudal vertebrae four, five, and six generally have longer rib-like processes projecting from their centra at right angles. On the first and seventh caudals these processes are exceptionally short. On the remainder they are further reduced or are entirely absent. Beginning with the eighth caudal vertebra secondary ventral processes, arising on the sides of the centra posteriorly, fuse at their tips to form a canal through which the caudal artery and vein pass. On the dorsal side each vertebra gives rise to a pair of lateral articulating surfaces (zygapophyses). Processes are reduced greatly or are lacking from the most caudal tail vertebrae whose centra are very slender.

APPENDICULAR SKELETON, figs. 13-15.

The pectoral girdle unites the forelimb, and the pelvic girdle unites the hindlimb, to the axial skeleton. These unions permit the turtle to bear the weight of head, neck, trunk, and tail upon the limbs when walking.

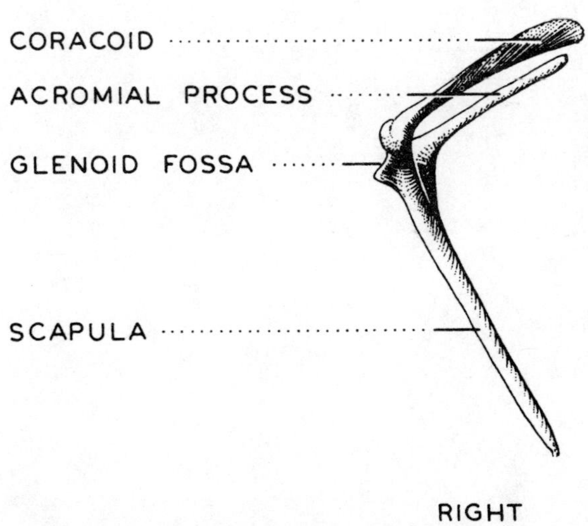

Fig. 14–Pectoral girdle, posterior

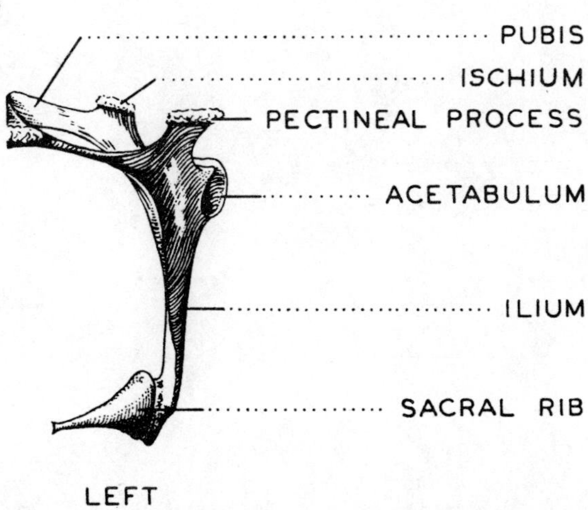

Fig. 15–Pelvic girdle, anterior

A. PECTORAL GIRDLE AND FORELIMB. The pectoral girdle of turtles lies between the carapace and the plastron and is attached via the scapula to the vertebral column.

1. SCAPULA. The longest, straightest bone of the pectoral girdle – also the most dorsal one. It is united proximally by a small cartilage to the first dorsal vertebra being attached at the angle between the first dorsal vertebra and the first rib. At the distal end of the scapula find the acromial process (pre-coracoid) projecting anteriorly at right angles from the scapula.

2. CORACOID, the third part of the pectoral girdle. It also lies at right angles to the scapula but is not fused to it as is the acromial process. The coracoids are joined by cartilage at their medial margins. The distal end of each coracoid is united by a ligament to the broad flattened end of the acromial process.

3. GLENOID FOSSA. A cavity formed by the union of scapula, acromion, and coracoid. It receives the head of the humerus to make the shoulder joint.

4. HUMERUS, the bone of the arm. In addition to its head note the two proximal tuberosities separated by a groove, and the neck below the head, and the heavy curved shaft expanded distally for the articulation of the foreleg bones, the *radius* and the *ulna.*

5. RADIUS, the smaller of the foreleg bones. It is about the same length as the ulna but is more slender.

6. ULNA, the larger bone of the foreleg. Determine which of these bones, radius and ulna, is medial and which is lateral in position.

7. CARPALS, the wrist bones as a group. Distally the ulna articulates with two bones, a lateral *ulnare* bearing a small knob on its side and a medially placed *intermedian* (intermediate) which lies between the ends of the radius and ulna. Distal to the radius is the *radiale* which may be fused to the bone lying medial to it – the *centrale.*

8. CARPUS. A carpal bone appears at the base of each *metacarpal* or bone of the forefoot (hand or manus).

9. PHALANGES, the bones of the toes (digits). There are two phalanges each in the first and fifth toes. All the others have three phalanges apiece. In each toe the distal phalanx bone forms a claw which is capped with a *horny* covering.

B. PELVIC GIRDLE AND HINDLIMB. Like the pectoral girdle, the pelvic, too, is composed of three pairs of bones. These are more firmly united than in the case of the pectoral girdle. The pelvic girdle is also housed within the turtle "shell" betwen the carapace and the plastron posteriorly.

1. ILIUM, the most dorsal bone of the pelvic girdle. Each ilium serves to unite the pelvic girdle to the vertebral column. Each is a stout bone, flattened proximally where it fuses with the pyramidal-shaped end of the first sacral rib on either side of the first sacral vertebra.

2. PUBIS, antero-ventrally placed and medially fused with its mate of the opposite side. The *pubes* form a triangular horizontal shelf, the apex of which is cartilaginous and is directed cephalad. This portion is sometimes termed the *epipubis.* Laterally each pubis exhibits a large *pectineal process* for attachment of muscles. Postero-laterally each pubis is fused to an ilium as well as to an *ischium* to form the *acetabulum socket* in which the head of the femur articulates thus forming the hip joint.

3. ISCHIUM, the third bone of the pelvic girdle. The *ischia* also fuse together in the midline and unite with the fused ilia anteriorly. On either side of this union is a large round hole, the *obturator foramen.* Each ischium exhibits a tuberosity projecting caudad. These processes are for the attachment of muscles.

4. FEMUR, the *thigh* bone. It resembles the humerus, having similar tuberosities and a somewhat larger head.

5. TIBIA AND FIBULA, the bones of the leg. The stronger one is the tibia – its proximal

end is club-shaped but this is not true of the fibula which is more slender throughout.

6. TARSALS, the ankle bones are six in number. Considerable fusion of more numerous embryonic bones has occurred here.

7. METATARSALS, the five bones of the foot – one for each digit.

8. PHALANGES, the bones of the digits. Like those of the forelimb there are two each in the first and fifth but three each for the other digits. The distal phalanx of the little toe is very small, being included within the web of the foot, therefore, there is no claw on the little toe.

CHAPTER III

Muscular System

The muscular system, as studied in anatomy, includes only the somatic or striated skeletal muscles. Heart muscle and visceral muscle belong largely to the circulatory and digestive systems respectively. The skeletal muscles of the turtle have not been as thoroughly studied as have those of some vertebrates, however, most muscles presented in this chapter can be readily identified with the aid of the table of origins, insertions, and functions together with the labeled drawings. This method, it is believed, will permit the identification and the learning of more muscles in a given amount of time than is possible with the ordinary style of laboratory manual. The student should be prepared to pass one or more oral examinations on his dissection of this system.

STRUCTURE OF A TYPICAL MUSCLE.

Most of the named muscles of the body wall (soma) consist of more or less spindle-shaped bundles of microscopic fibrils. These bundles have a thicker middle portion, the *belly*, and two rather slender ends, one of which, the *origin* is more or less immovable and the other, the *insertion* which is more or less movable. White glistening *tendons* generally attach a muscle at each of its ends to a bone or to a *ligament.* Ligaments hold bones together, their fibers blending securely with the fibrous *periosteum* which covers the bones. Since tendons generally attach muscles to bones their fibers blend with the periosteum at one end and with the muscle fibers at the other.

Along their sides adjacent muscles are held together by connective tissue or *intermuscular fascia. Subcutaneous fascia* unites superficial muscles with the overlying skin. Broad sheets of connective tissue covering flat muscles such as the oblique and transverse abdominals, in most higher vertebrates, are termed *aponeuroses.* These muscles are not very evident in turtles owing to their being firmly attached to the plastron.

Care must be exercised in "dissecting" muscles for one must *avoid cutting them* but rather they must be separated by tearing away the intermuscular fascia by means of the back side of a scalpel blade, the scalpel handle, or a strong probe or seeker. Every muscle identified should be thus fully separated from its fellows and its origin and insertion identified by means of the tables and diagrams. Be sure to break away the carapace wherever it interferes with your progress.

If at any time it becomes necessary to cut or to remove a muscle in order to identify others beneath it, you should cut across the belly of the muscle so that it can later be studied by merely bringing together the cut ends.

PREPARATION OF THE TURTLE FOR THE STUDY OF MUSCLES.

The muscles of the head, neck, limbs, and tail can readily be exposed merely by skinning these parts. In order to study the more proximal limb, neck, and tail muscles, however, as well as those of the trunk it will be necessary to remove the plastron. If this has not been previously done you must cut through each of the bridges uniting the carapace and plastron. Next cut through the skin where it joins the plastron and the latter can now be easily removed. A bone saw, hack saw blade, or strong bone shears will suffice for this job of bone cutting. CAUTION: The muscles adhering to the dorsal side of the plastron should be cut off close to the plastron in this case.

IDENTIFICATION OF MUSCLES.

Rather than reading a detailed description of each muscle it will be possible to save much time

by referring to the tables of the muscles given below. These may be used in connection with the illustrations so that the origin, insertion, and action of each muscle may be learned as it is identified. Figs. 16-20.

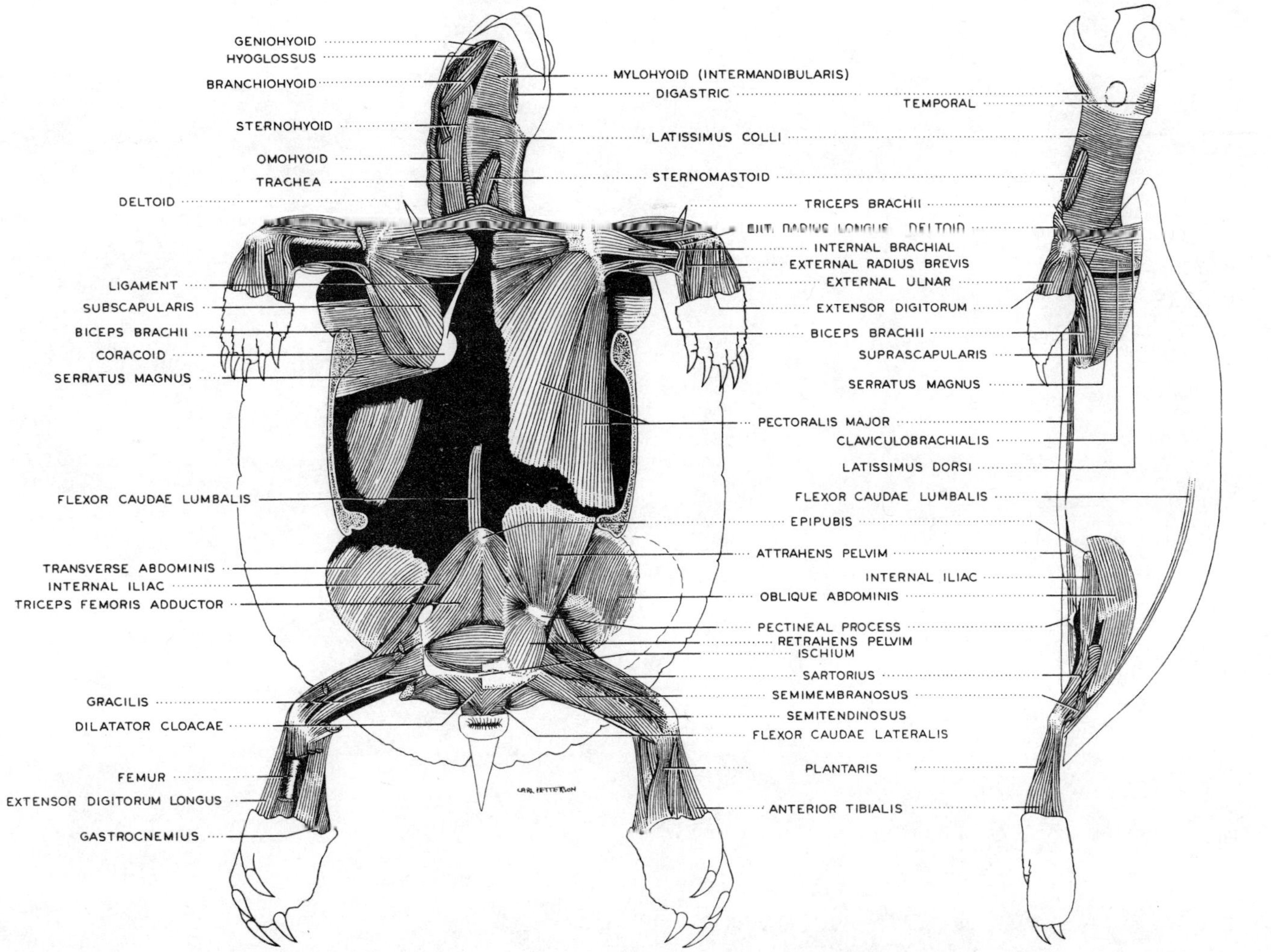

Fig. 16—Muscles, ventral

Fig. 17—Muscles, lateral

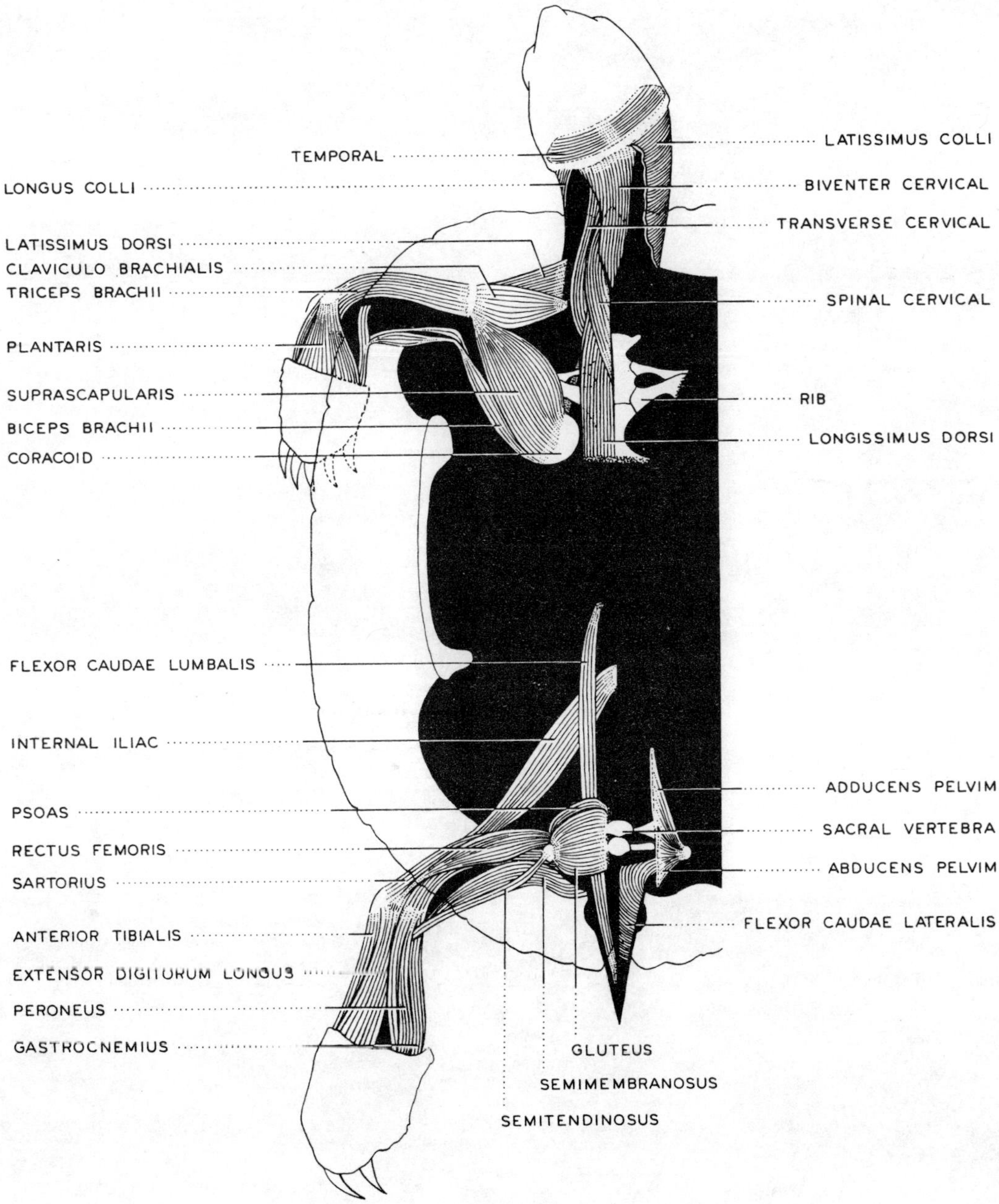

Fig. 18—Pelvic and pectoral muscles, ventral

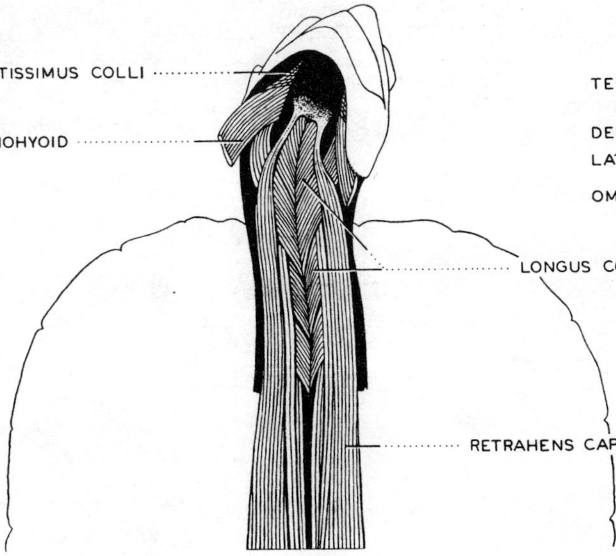

Fig. 19—Deep neck muscles, ventral

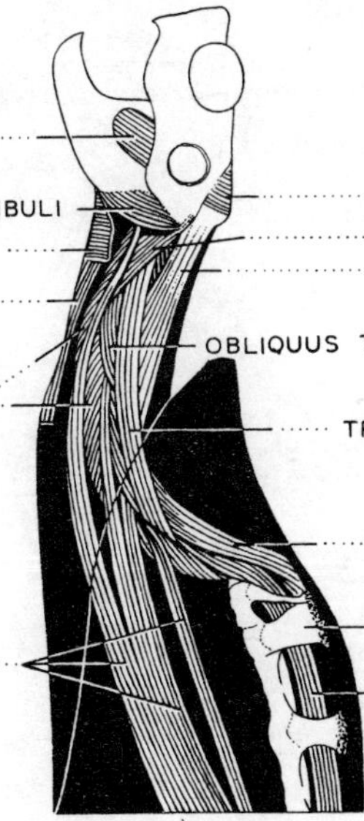

Fig. 20—Deep neck and shoulder muscles, left lateral view

TABLE I. SKELETAL MUSCLES OF THE TURTLE

MUSCLE	ORIGIN	INSERTION	ACTION
HEAD AND NECK MUSCLES			
Temporal	Supratemporal fascia	Between rami of jaws	Elevates jaw (closes mouth)
Digastric	Ventral side of bulla	Ventral ramus of jaw	Depresses jaw (opens mouth)
Cervico-capitis	Ant. cervical vertebrae	Posterior tip of bulla	Turns head (side-to-side)
Sternomastoid	Ligament at tip of posterior hyoid	Base of skull	Turns head, supports pectoral girdle
Omohyoid	Anterior margin of coracoid	Edge of ant. hyoid	Helps depress jaw, retract head, and lower oral floor
Latissimus colli	Sides of cervical vertebrae	With mate on midventral line	Compresses throat as in swallowing
Longus colli	Cervical vertebrae	Cervical vertebrae	Extension of neck
Retrahens capitis collique	Cervical vertebrae and roof of mouth	Carapace ant. to ilium, and dorsal vertebrae	Retraction of head and neck
Longissimus dorsi	Vertebral column dorsal to ribs	Posterior cervical vertebrae	Extension of neck
Transverse cervical	First cervical vertebrae	Posterior cervical vertebrae	Arches neck (flexion)
Biventer cervical	Middle cervical vertebrae	Fascia over skull and temporal muscle	Elevation of head
Spinal cervical	Last cervical vertebrae	Middle cervical vertebrae	Extends and elevates neck
Obliquus transversalis colli	Cervical vertebrae	Cervical vertebrae	Twists and extends neck
PECTORAL GIRDLE AND FORELIMB MUSCLES			
Pectoralis major	Precoracoid, distally and plastron	Head of humerus	Adduction of humerus posteriorly
Deltoid	Anterior margin of plastron & precoracoid ventrally	Suprascapularis muscle anteriorly & head of humerus	Anterior and lateral extension of arm
Suprascapularis	Coracoid, ventrally	Head of humerus	Flexion of arm
Biceps brachii	Posterior margin of coracoid	Radius and ulna	Forearm flexion & adduction
Triceps brachii	Head of humerus & periosteum over scapulo-precoracoid joint	Ulna	Extension & rotation of forearm
Internal brachial	Head of humerous	Radius and ulna	Forearm flexion
External radius brevis	Humerus, distally	Carpals	Extends and rotates hand outwards
External ulnar	Humerus, distally	Carpals	Extends and rotates hand outwards
External radius longus	Alongside humerus and radius	Carpal and phalanx of first digit	Extension of hand
Extensor digitorum	Humerus, distally	Metacarpals & phalanges	Extension of hand
Serratus magnus	Carapace, anteriorly and edge of bridge	Coracoid	Anterolateral extension of shoulder & foreleg
Palmaris	Distal end of humerus	Broad ligament continuing as phalangeal fascia	Flexion of hand & digits
Suprascapularis	Coracoid, dorsally	Head of humerus and claviculobrachialis	Posterior adduction and flexion of arm
Claviculobrachialis	Along scapula	Head of humerus	Ventro-lateral extension of arm
Latissimus dorsi	Carapace, ventrally and along scapula	Neck of humerus via a broad ligament	Lifts arm dorsally & anteriorly

TABLE I. SKELETAL MUSCLES OF THE TURTLE — CONTINUED

MUSCLE	ORIGIN	INSERTION	ACTION
MUSCLES OF PELVIC REGION			
Attrahens pelvim	Pectineal or lateral pubic process	Plastron, posteriorly	Fixation of pelvic girdle
Oblique abdominis; Transverse abdom.	(These are mostly removed along with the plastron to which they are attached)		
Triceps femoris adductor	Pubis, ventrally & ischium	Proximal head of femur	Extends thigh
Internal iliac	Dorsal surface, margin & apex of pubis	Distal to head on femur	Flexes leg
Flexor caudae lumbalis	Periosteum above thoracic vertebrae 2/3 way back from neck	Ventral tail fascia	Pulls in tail
Flexor caudae lateralis	Head of ilium & carapace	Tail, mid-dorsally	Lateral flexion of tail
SKELETAL MUSCLES CONTINUED			
Dilator cloacae	Sacral vertebrae & along tail	Tail, midventrally & anal skin	Retracts & rotates tail
Sphincter cloacae	Sacral vertebrae & along tail	Tail, midventrally & anal skin	Retracts & rotates tail
Retrahens pelvim	Lateral pubic process and pelvic ligament	Plastron, posteriorly	Fixation of pelvic girdle
Adducens pelvim	Carapace near anterior head of ilium	Anterior head of ilium	Anchors pelvic girdle to carapace
Abducens pelvim	same as above	same as above	same as above
HINDLIMB MUSCLES			
Sartorius	Via tendon to pelvic ligament	Proximal end of tibia	Extension of leg
Semimembranosus	Posterior border of ischium & pelvic ligament	Multiple insertions on tibia	Flexes leg
Semitendinosus	Proximal end of ilium	Body of fibula	Flexes leg
Gracilis	Pelvic ligament via a long slender tendon	Proximal head of tibia, along inner surface	Flexion & slight rotation of leg
Anterior tibialis	Head & body of tibia	First metacarpal & other small insertions	Pulls foot towards body as in walking
Plantaris	Femur	By broad fascia over plantar area & by tendons to phalanges	Retracts claws, extends foot, flexes leg
Gluteus	Head of femur along edge of ilium & sacral vertebrae	Head of femur	Pulls leg forward & dorsally. Helps retract leg.
Vastus internus	Lateral pubic process & margin of femur	On tibia with sartorius	Aids sartorius in extension of leg
Rectus femoris	On ilium via a tendon	Metatarsus of last digit	Leg extension
Gastrocnemius	Distal end of femur	Metatarsus of last digit	Pulls foot to body, spreads claws, & helps pull leg posteriorly
Peroneus	On body of fibula	Metatarsus of 4th & phalanx of 5th digits	Spreads claws & pulls them posteriorly
Extensor digitorum longus	Distal end of femur	Metatarsus	Pulls claws together & lifts & extends toes
Crureus	Head of femur	Head of tibia with sartorius	Leg extension
SKELETAL MUSCLES CONCLUDED			
Semimembranosus	Head of ilium & caudal vertebrae	Ligaments of semimembranosus	Flexes leg
Psoas	Ilium & sacral vertebrae	Head of femur, ventrally	Pulls leg under carapace

CHAPTER IV

Digestive and Respiratory System

If by chance the muscular system has been omitted, then the plastron will now need to be removed. If the bony bridges joining plastron to carapace have not been previously cut, this should be done now as directed in the foregoing chapter.

A. PERITONEUM AND COELOM.

1. Parietal peritoneum, a tough thin membrane covering all visceral organs and forming an inner lining for the plastron and carapace, where they surround the visceral cavities. Muscle layers are absent or much reduced in the ventral body wall since the plastron fully supports this region in most turtles. The muscles of the girdles and limbs are present as usual.

2. Pericardial sac, a triangular sac continuous with the parietal peritoneum in the ventral midline just posterior to the pectoral girdle. This sac encloses the heart and is termed the *parietal pericardium* while the thin outer covering of the heart itself is the *visceral pericardium.*

3. Pericardial cavity, a portion of the original coelom lying between the parietal and visceral layers of the pericardium and surrounding the heart. Remove the ventral portion of the parietal pericardium thus exposing the heart but preserve the dorsal portion of the sac for future reference.

4. Ventral abdominal veins. These large vessels form a letter H-shaped design in the ventral peritoneum. Their posterior limbs arise from the pelvic girdle and their anterior limbs enter the right and left liver lobes, respectively. A branch from each anterior limb extends anterolaterally to the pectoral girdle on its own side. A cross anastomosis (transverse abdominal vein) will be noted midway between the apex of the pelvic girdle and the posterior borders of the liver lobes. Cut away the ventral peritoneum but leave the vessels intact. Now cut both ventral abdominals just posterior to the cross anastomosis and reflect the anterior limbs forward so that the abdominal viscera may be clearly seen.

5. Pleuroperitoneal cavity, the space exposed by removing the ventral parietal peritoneum. This space includes most of the visceral organs.

6. Coelom, the pericardial and the pleuroperitoneal cavities.

B. VISCERA AND MESENTERIES, figs. 21 and 22.

Break away the sides of the carapace (use bone shears or pliers) between the fore and hind legs to facilitate viewing the viscera. Various masses of yellowish-green *fat* and more peritoneum may be removed. The form, color, and relationships of the following organs should be learned.

1. Liver, a brown gland — the largest in the body. Its left lobe nearly covers the stomach and extends to the right, dorsal to the ventricle of the heart. It expands into the right and largest lobe which covers part of the small and large intestines. An extension of the left lobe lies dorsally in the lesser curvature of the stomach. If the specimen is a female, ovaries containing eggs of various sizes and of spherical shape will be seen lying just posterior to the liver and against the dorsal body wall.

2. Hepatic mesenteries, in which the anterior limbs of the ventral abdominal veins are contained, are short mesenteries uniting the right and left liver lobes to the ventral peritoneum. These mesenteries correspond to the falciform ligament seen in most vertebrates.

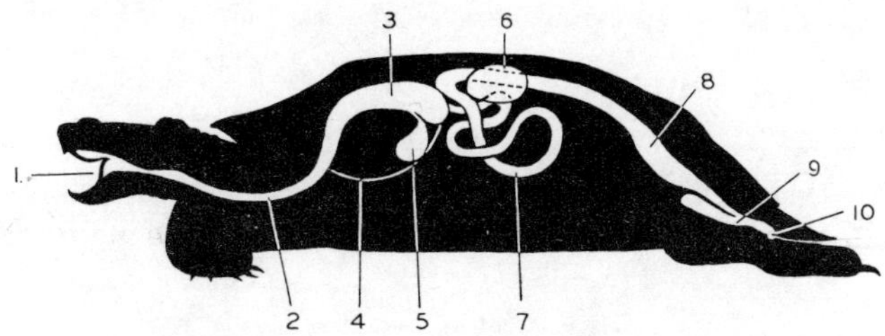

Fig. 21—Digestive system, lateral view (numbers corresp. to those in Fig. 22.

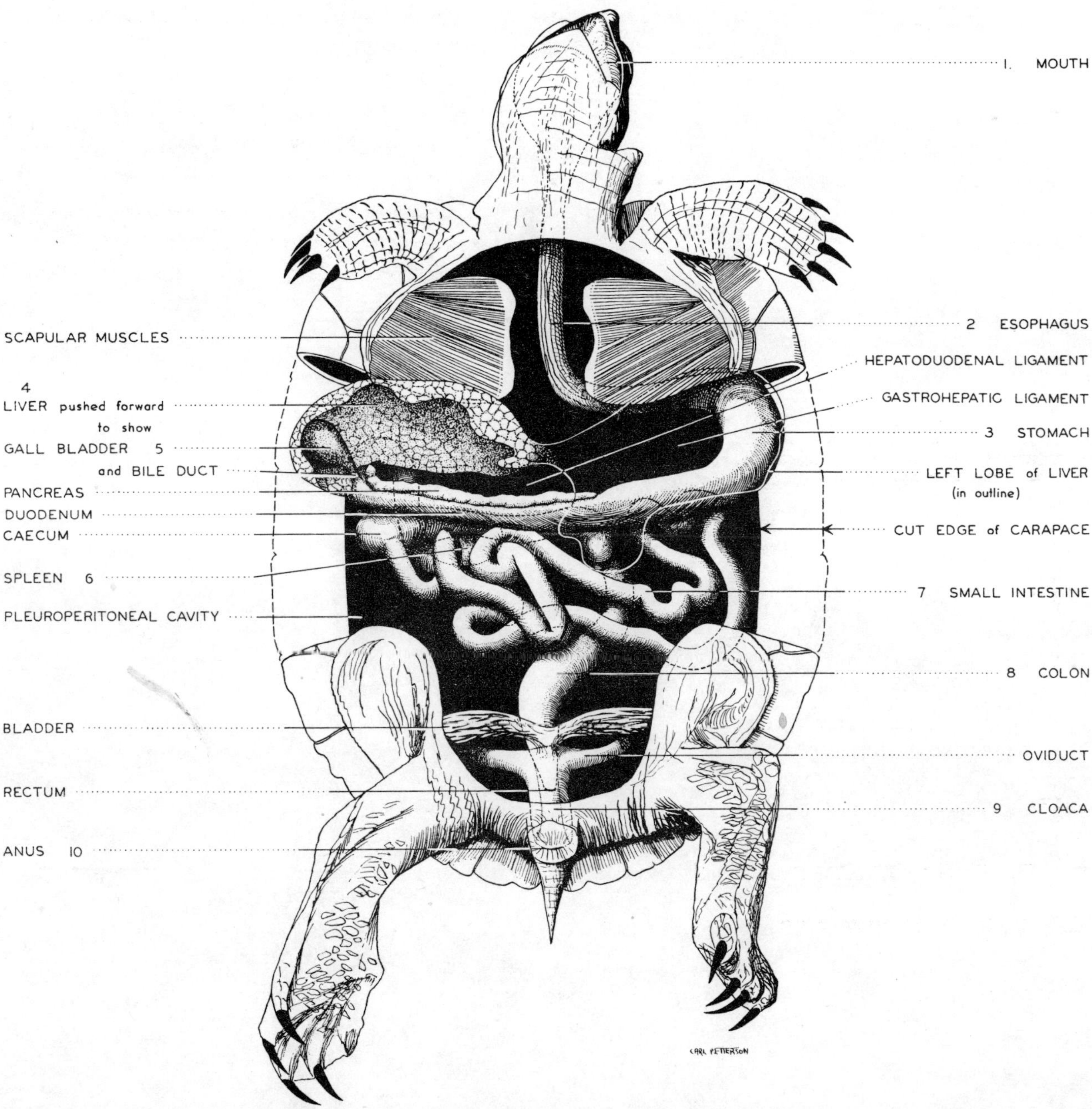

Fig. 22—Digestive system, ventral

3. Transverse septum, a membrane formed of the parietal peritoneum dorsally and of the underlying dorsal wall of the pericardium. It lies in an oblique plane owing to the more caudal position of the heart in turtles (the heart lies farther anterior in fishes and in amphibians).

4. Coronary ligament, an attachment of the liver to the transverse septum. Follow the parietal peritoneum as far anteriorly as possible and observe its continuity with the dorsal portion which forms the inner lining of the carapace. Note also the connections between the ventral and dorsal peritoneum in the region of the pelvic girdle.

5. Esophagus (gullet), a slender tube uniting the pharynx with the stomach. It may be seen in part by lifting the left liver lobe far over to the right. It joins the stomach at a point just dorsal to the left atrium of the heart.

6. Stomach, a curved tubular digestive organ leading into the small intestine. The stomach is much greater in diameter than either the esophagus or the intestine.

7. Gastrohepatic ligament, a short mesentery attaching the stomach all along its lesser curvature to the left lobe of the liver.

8. Duodenum, the first part of the small intestine. It begins where the stomach ends just dorsal to the bridge of liver which connects the left and right liver lobes.

9. Hepatoduodenal ligament, a mesentery uniting the duodenum with the right hepatic lobe along the middle of its dorsal surface.

10. Pancreas, a long slender pale gland lying along the duodenum at the attachment of the hepatoduodenal ligament.

11. Pancreatic duct. This is fairly stout and passes from the pancreas into the duodenum about one half centimeter from the right end of the pancreas. Tease away the pancreas at this point and identify this duct.

12. Gall bladder, the bile reservoir. It is greenish colored and lies on the dorsal side of the right liver lobe near its right border.

13. Bile duct (ductus choledochus), a short stout duct which empties the gall bladder into the duodenum. It can be easily identified and traced to its termination.

14. Small intestine proper, that part of the small intestine beyond the duodenum, beginning at the sharp curve posterior to the right border of the right liver lobe. It forms several coils and terminates by empting into the beginning of the large intestine, the caecum. The right ovary may now be removed, if necessary, to facilitate tracing the small intestine.

15. Dorsal mesentery (mesentery proper). This attaches the intestine to the dorsal midline of the body cavity. Lift the intestine enough to demonstrate this mesentery.

16. Mesoduodenum. This is fused to the hepatoduodenal ligament so that the two are indistinguishable. Both of these structures are mesenteries.

17. Mesogaster, a separate mesentery distinguishable from the gastrohepatic ligament. It connects the greater curvature of the stomach with the dorsal midline of the body cavity.

18. Colon, the large intestine. Trace the small intestine to its junction with the caecum (beginning of large intestine) located in the posterior right quadrant of the visceral cavity.

19. Mesocolon, the mesentery of the colon.

20. Cloaca, the terminal portion of the alimentary canal. It opens to the exterior via the *anus*. Cut out the anterior mid-portion of the pelvic girdle, using bone shears, and trace the colon, noting its ascending, transverse, and descending limbs, to the cloaca. The transverse colon will be found to be fused to the mesogaster. A thin-walled bilobed *urinary bladder* will be seen just dorsal to the pelvic girdle.

21. Spleen, a spheroid vascular gland whose functions include destruction of old red blood cells, salvaging hemoglobin, the manufacture of certain kinds of white blood cells, and other specialized duties. It lies in the mesocolon just anterior to the caecum but is dorsal in position, hence careful search in the mesocolon may be required in order to find it.

C. ORAL CAVITY AND PHARYNX, fig. 23.

Force open the jaws and cut the soft tissue back through the angles of the jaws until the mouth can be opened enough to see the beginning of the esophagus. A probe can now be passed through the entire esophagus into the stomach. Lifting the stomach over toward the right side will facilitate passage of the probe.

1. Oral (buccal) cavity, the space enclosed by the jaws. Note the *horny beaks* which cover the jaws. Observe the relationship between this horn and the skin of the nostril, face, and neck.

2. Beak. True teeth are lacking in turtles but the horny beaks extend into the mouth and develop tooth-like ridges paralleling the edges of the beaks. Compare these for upper and lower jaws.

3. Anterior nares (nostrils), two small openings in the tip of the snout. These lead back through the nasal passages which open on the roof of the mouth as the:

4. Posterior nares. These can best be found by passing a slender probe back through the nostrils and nasal passages where their openings will

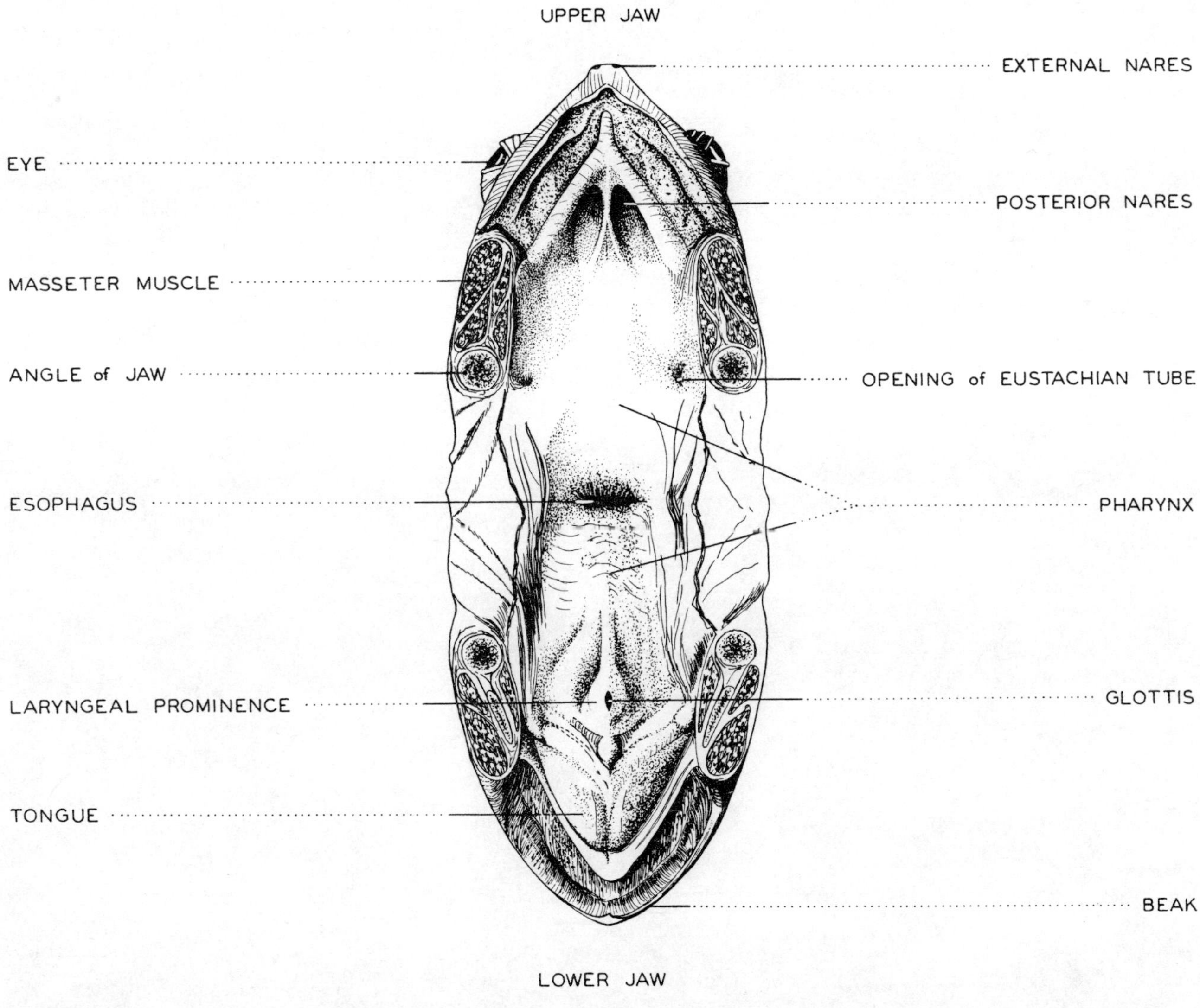

Fig. 23–Mouth and Pharynx

be seen just posterior to the bony plate of the upper beak and in a depression bounded anterolaterally by a Y-shaped ridge of soft tissue.

5. Tongue, a triangular papillated fleshy structure in the floor of the buccal cavity which is firmly attached along its entire length.

6. Pharynx, a cavity extending from the posterior border of the tongue (its base) to the beginning of the esophagus.

7. Laryngeal elevation or prominence, a raised portion of the floor of the pharynx just behind the base of the tongue.

8. Glottis, a slit-like opening in the center of the laryngeal elevation. It opens into the simple larynx.

9. Auditory (Eustachian) tube, a small canal leading from the roof of the pharynx to the *middle ear cavity*. Both tube and cavity are derivatives of the first pharyngeal pouch. (The openings into the auditory tube may have been damaged when cutting back through the angles of the jaws.) Find the slit-like opening of one of these tubes and probe it to the cavity of the middle ear. A disc of skin (tympanum) covering this cavity may thus be identified. Cut away this disc and view the cavity beneath it.

D. RESPIRATORY DUCTS AND LUNGS, figs. 24 and 25.

1. Larynx. The upper end of the respiratory duct. It leads into the trachea below and opens above into the pharynx via the glottis. Find a pair of *arytenoid cartilages*, one on either side of the larynx. These support wedge-shaped flaps on either side of the glottis and serve to regulate the size of this opening. A ring-shaped *cricoid cartilage* lies posterior to the glottis. It is wide ventrally but narrow dorsally.

2. Trachea, a connection of the larynx with the bronchial tubes which enter the lungs. Slit open the glottis and anterior part of the trachea along the mid-ventral line and notice that the glottis opens into a spheroid chamber anterior to the first cartilaginous ring of the trachea — this is the cavity of the larynx. Now slit open the skin of the neck along the ventral midline. Spread the skin and muscles apart and trace the trachea to its bifurcation into the *right and left bronchi.*

3. Bronchus, a branch formed by the bifurcation of the trachea. There are two main branches, a right and a left. With a curved flexible probe follow the left bronchus as far as you can. Reflect the stomach and large intestine to the right, open the peritoneum over your probe and note the lungs.

4. Lungs, reddish spongy respiratory organs lying against the carapace on each side just posterior to the pectoral girdle. Open the left lung by a lateral incision passing to the left from your probe. Study the texture of the lung, observing its air pockets or *alveoli* and the tough connective tissue strands which separate them. Note the relationship between peritoneum and lung in both its anterior and posterior regions. Is the lung entirely retroperitoneal or does part of it extend into the pleuroperitoneal cavity?

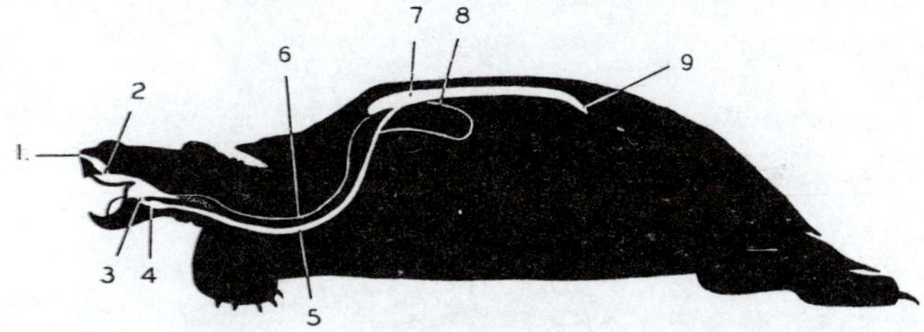

Fig. 24—Respiratory systems, lateral view (nos. as those in Fig. 25).

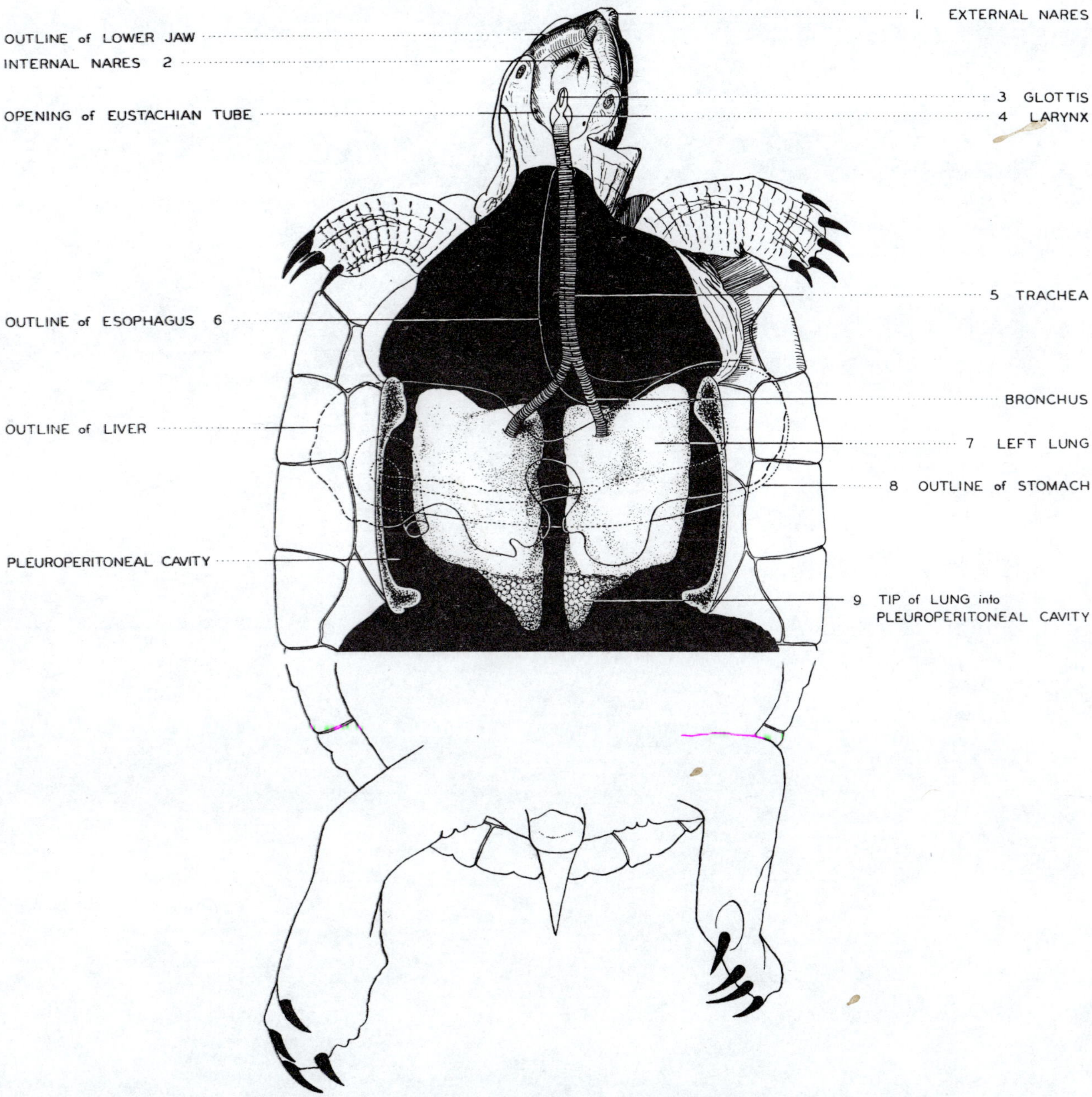

Fig. 25—Respiratory system, ventral

CHAPTER V

Circulatory System

The heart, arteries, and veins are the gross features of the circulatory system and they should be dissected with great care and learned as completely as possible. Do not cut any sizeable vessel unless directed to do so. Follow directions carefully and preserve all possible structures for subsequent study.

A. THE HEART, figs. 26-29.

The heart lies within a pericardial sac the ventral wall of which should have been removed in the study of the digestive system and mesenteries. The transversely expanded ventricle whose bluntly rounded apex points caudally is the most conspicuous part, but dorsal to this lie the two atria separated by an interatrial septum and the sinus venosus which lies dorsal to the atria.

1. Ventricle, the main muscular pump of the heart. It sends blood to all parts of the body including the lungs. Sever the thin ligament attaching the apex of the ventricle to the posterior pericardial wall (this is probably a remnant of the embryonic ventral mesentery), lift the apex cephalad and note the connections between ventricle, atria, and sinus venosus. Return the ventricle to its place and notice the large caval veins entering the sinus and the large arteries leaving the anterior side (base) of the ventricle.

2. Right atrium. This receives venous blood from the sinus venosus and empties it into the ventricle via the *right atrio-ventricular valve.*

3. Left atrium. It receives arterial blood via the pulmonary veins from each lung and empties

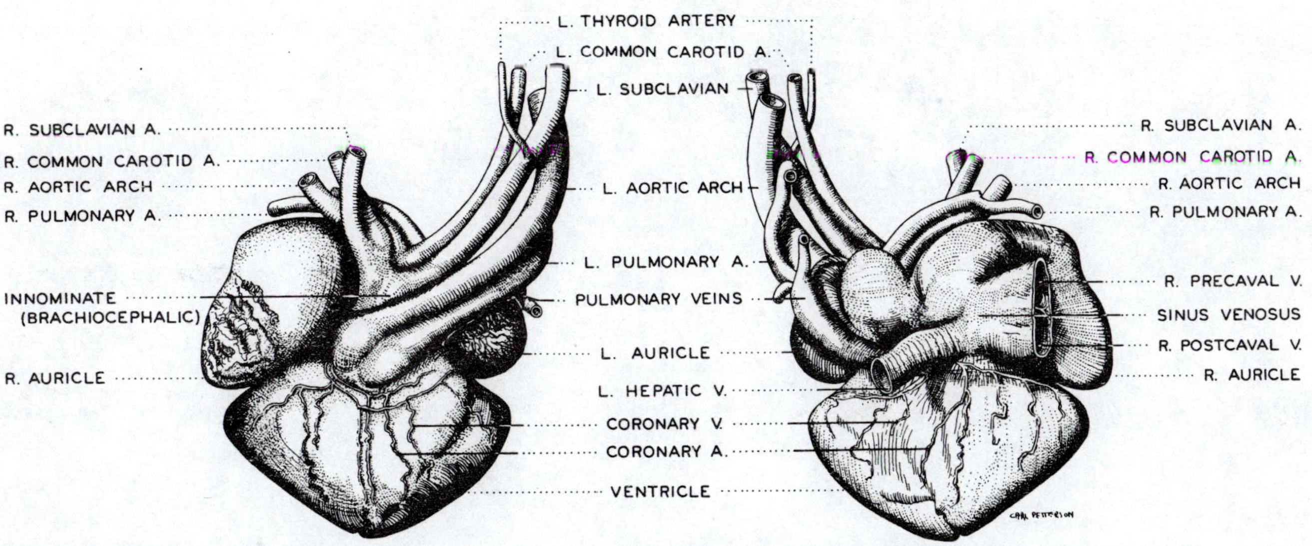

Fig. 26—Heart, ventral

Fig. 27—Heart, dorsal

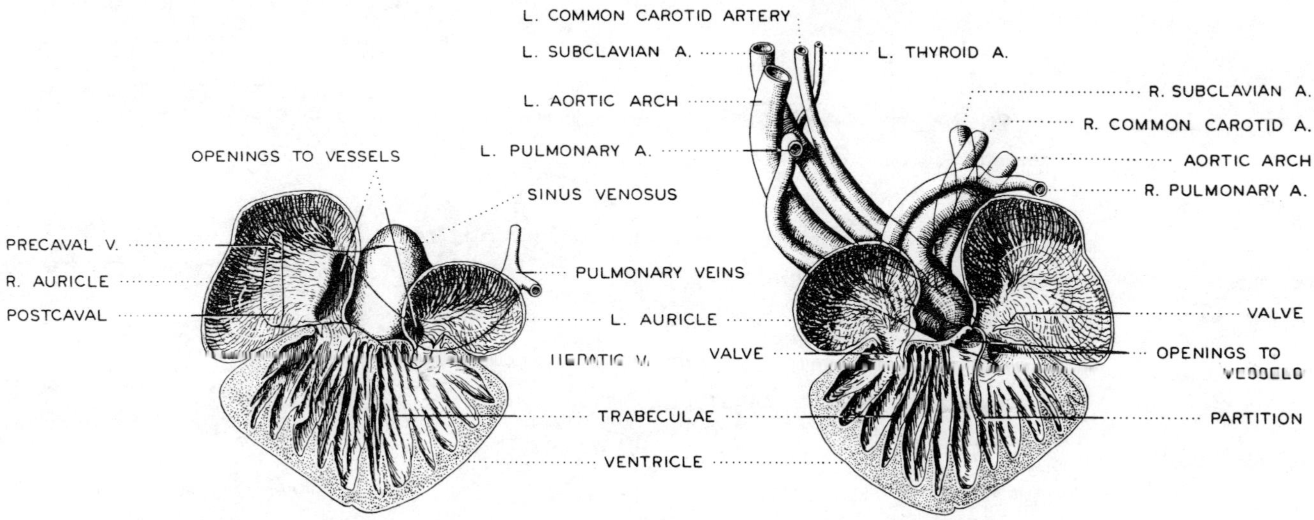

Fig. 28—Heart, internal, dorsal half

Fig. 29—Heart, internal, ventral half

it into the ventricle via the *left atrio-ventricular valve* where it mixes but slightly with the systemic venous blood which largely goes to the lungs for oxygenation. Further mixing of arterial and venous blood is avoided by the spongy interventricular septum.

4. Sinus venosus. It receives blood from the systemic veins of the entire body excepting the lungs. Four large veins, to be described below, empty their contents into the sinus.

B. VENOUS TRUNKS OF THE HEART, figs. 30 and 31.

In addition to the two pulmonary veins which empty into the left atrium there are four large veins supplying venous blood to the sinus. Displace the ventricle cephalad to expose the sinus and identify the following veins:

1. Left precaval (anterior vena cava), swings around the border of the left atrium and enters the sinus.

2. Left hepatic. It drains the left liver lobe from which it appears in the bridge of liver and terminates at the left angle of the sinus venosus.

3. Postcaval (posterior vena cava), a very large vessel draining the body caudal to the heart. It emerges from the right liver lobe and enters the right angle of the sinus.

4. Right precaval (anterior vena cava). It enters the sinus just anterior to the entrance of the postcaval. It may best be seen by pressing the heart toward the left. The precavals drain blood from the anterior regions of the body.

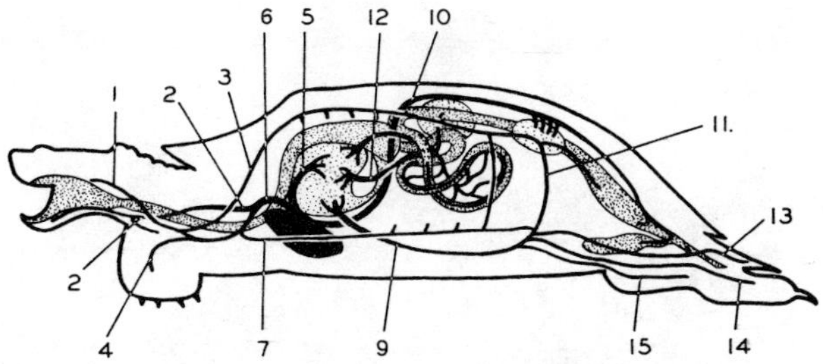

Fig. 30—Venous system, lateral view, (nos. as those in Fig. 31).

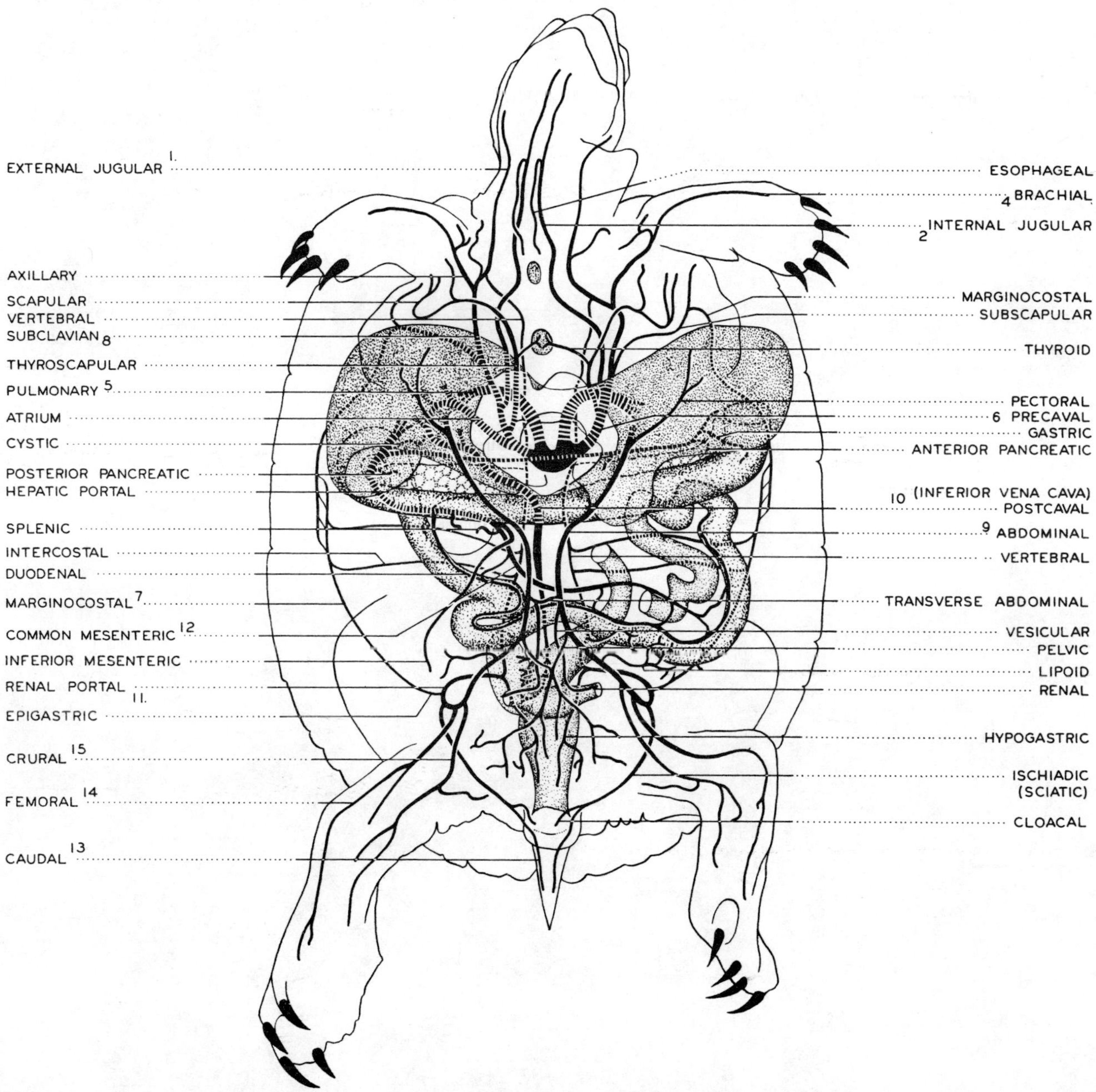

Fig. 31—Venous syst., ventral

C. ABDOMINAL AND RENAL PORTAL VEINS, figs. 31, 31a, and 32.

Several different arrangements may be found in the venous system, particularly among the veins tributary to the abdominals.

1. Abdominals, a pair of vessels arising on either side of the pelvic girdle and passing cephalad to disappear into the liver lobes. A cross anastomosis sometimes termed *transverse abdominal vein* will be seen just in front of the pelvic girdle. These veins were noted under the digestive system.

2. Pectoral veins. These veins arise in the pectoral girdles and drain into the abdominals at points near the liver, one on either side.

3. Pericardial veins. Each abdominal vein receives one of these from the pericardial sac shortly before entering the liver.

4. Vesicle vein, a small vein which arises from the bladder and unites with the abdominal vein of its own side just dorsal to the transverse abdominal. Since the tributaries of the ventral veins are all paired, it will be necessary to describe only those on the left side from here on. If the veins on the right are better injected than those on the left they may be followed instead.

5. Pelvic vein. It drains the pelvic girdle into the abdominal vein which it joins midway between the apex and the base of the girdle. It passes ventrally over the pelvic muscles. One pelvic vein may be much larger than the other.

Now limber up the hind leg on the side you are following by alternate flexion and extension, abduction and rotation until the leg can be held towards the ventral midline. Cut through the skin next to the carapace, from the thigh back to the tail, being careful to spare important blood vessels. Remove the skin from the leg and tail and trace the abdominal vein laterally along the base of the leg.

6. Crural vein. One which arises from medial thigh muscles and unites with the abdominal just caudal of the pelvic vein (or it may unite with the femoral just before reaching the abdominal or it may unite with the external iliac from which a connection is made to the abdominal).

7. Lipoidal or adipose vein. It arises from fat in the groin and enters the abdominal just caudal of, or beside, the crural. It may be larger than the crural.

8. Femoral, a large vein from the dorso-lateral side of the thigh. It enters the abdominal about one and one half inches beyond the union of the crural and the abdominal. Trace the femoral vein into the leg by separating the muscles.

9. External iliac, a short vein extending from the union of abdominal and femoral veins to the junction of the renal portal and epigastric veins, figs. 31 and 31a. The external iliac receives several small veins from the abdominal wall.

10. Epigastric vein. It parallels the epigastric artery along the curved edge of the carapace and enters the external iliac vein (or it may enter the portal) at a point where the latter turns at right angles posteriorly to run between the carapace and the upper thigh. Clear away loose tissue and break off the carapace all around the postero-lateral margin (use bone shears or pliers).

11. Ischiadic vein, a tributary to the renal portal vein. The ischiadic receives the two following veins as tributaries.

12. Caudal vein, a continuation of the ischiadic along the side of the tail.

13. Cloacal vein. It drains the cloaca and anal region and unites with the caudal thus forming the ischiadic.

14. Renal portal. Find the vein at the junction of the epigastric and external iliac veins and trace it anterodorsally where it penetrates the pleuroperitoneum and enters the kidney, the latter lying in a retroperitoneal position, i.e. dorsal to the pleuroperitoneum. It will be necessary to turn the viscera cephalad and to cut through the pleuroperitoneum transversely at a point midway between heart and pelvic girdle so as to carry the left lung forward in order to obtain a clear view of the kidney which lies against the carapace postero-dorsally and near the midline. Remove the

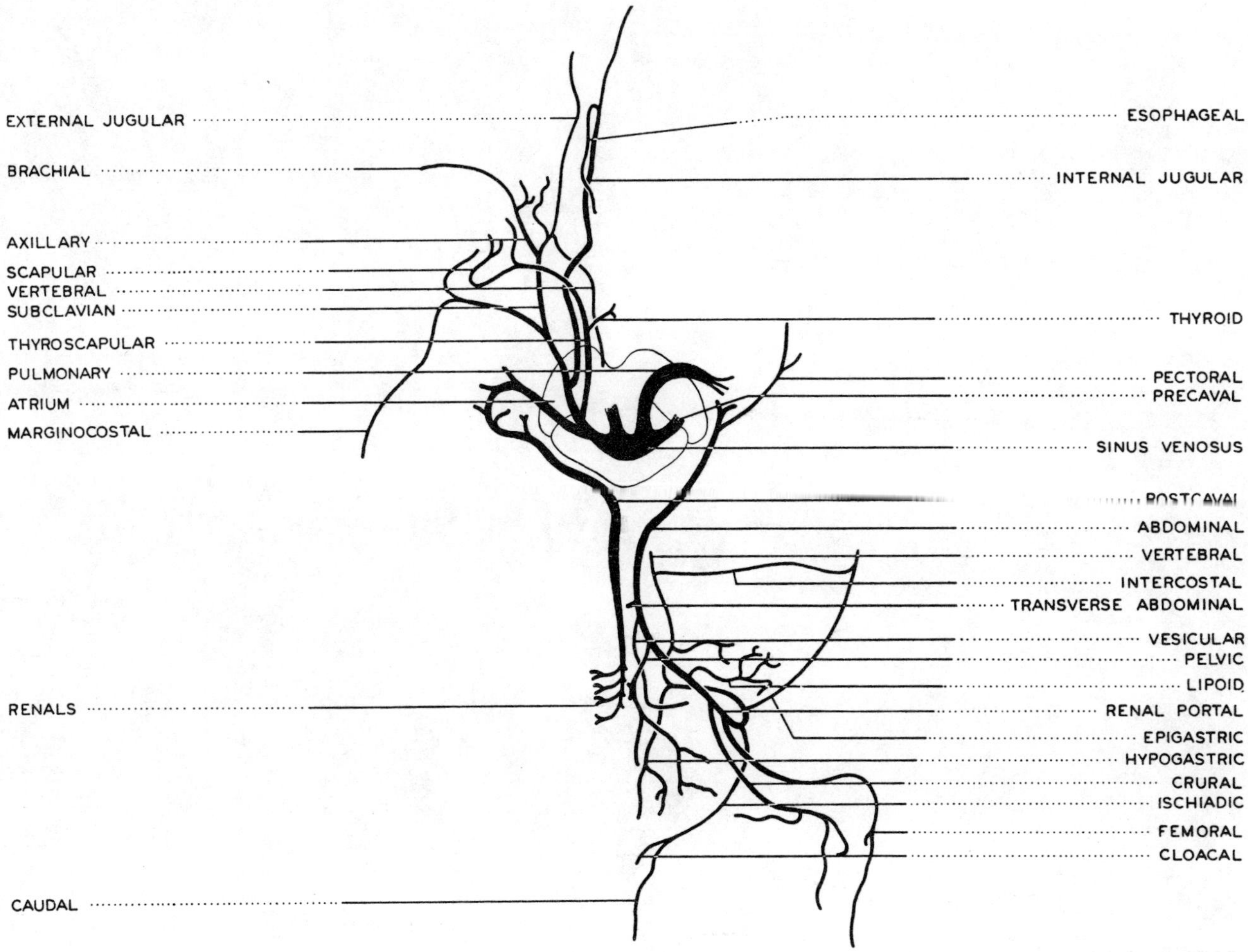

Fig. 31A—Variation from typical venous system seen in Fig. 31.

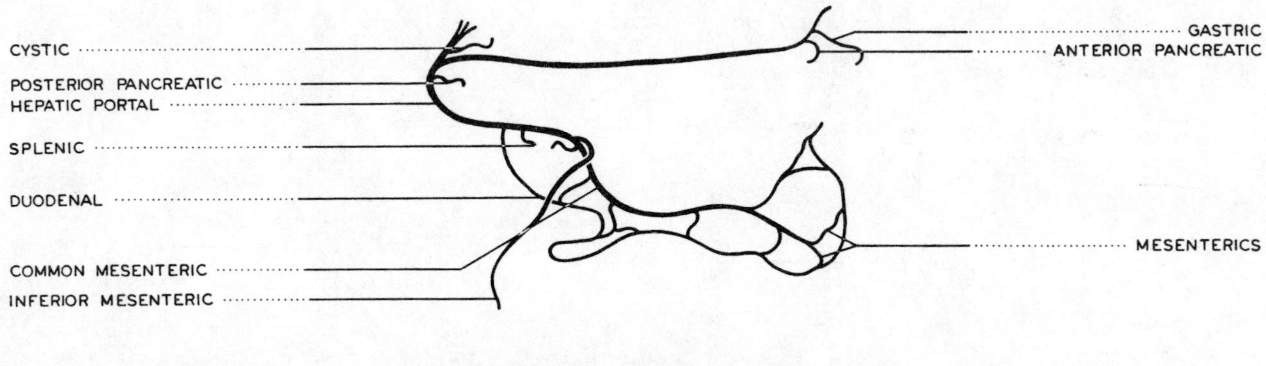

Fig. 31B—Scheme of hepatic portal syst. (refers to Figs. 33 & 34).

pleuroperitoneum covering the kidney so as to be able to see all vessels in this region. The renal portal may also receive small veins from the carapace and from the muscle mass just posterior to the kidney. Veins from the carapace may also enter the kidney directly on its dorsal surface.

15. Vertebral vein, the anterior one of two branches from the renal portal at its point of contact with the kidney. The vertebral passes cephalad dorsal to the arches of the ribs which fuse with the carapace a short distance lateral to the vertebral column.

16. Intercostal veins. These enter the vertebrals, on either side, meeting them at the intercostal spaces. They are sometimes injected poorly or not at all.

17. Marginocostal vein, the anterior continuation of the epigastric vein. It serves to connect the intercostal veins peripherally along the margin of the carapace. The marginocostal also makes connection with the abdominal veins some distance caudal to the heart.

18. Hypogastric vein, the posterior branch of the renal portal. It passes caudad over the ventral side of the kidney to the bladder while receiving branches from the rectum, cloaca, and male genitalia.

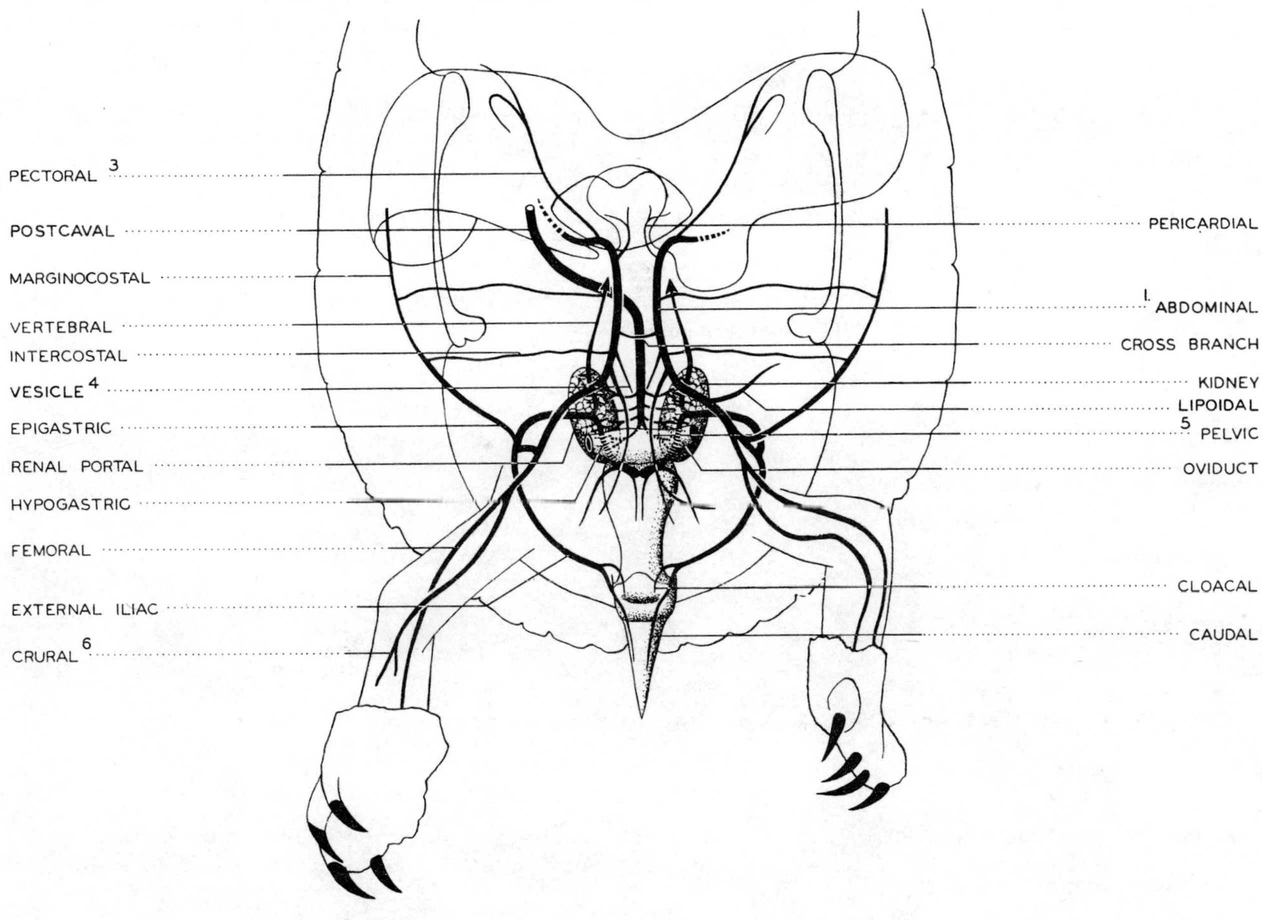

Fig. 32—Veins in renal portal or abdominal regions

D. HEPATIC PORTAL SYSTEM, figs. 33 and 34.

This system of veins carries venous blood from stomach, small intestine, pancreas, spleen, and gall bladder into the liver where many important functions are performed such as glycogen storage, removal of certain blood pigments, and detoxification of the blood. Identify and learn the following veins of the hepatic portal system:

1. Hepatic portal. It passes from left to right along the dorsal side of the liver lobes. It sends many branches into the liver tissue and receives the right and left ventral abdominal veins as tributaries in addition to the branches listed below.

2. Gastric veins. They enter the left portion of the hepatic portal from the stomach and are small but numerous—at least five or six may be readily seen.

3. Anterior pancreatic veins. Two or three of these may be seen passing from the pancreas to the hepatic portal vein just to the right of the median bridge of liver tissue.

4. Cystic veins, tributary to the hepatic portal from the bile duct and from the gall bladder. Since these are quite small they may not be injected and thus they may be difficult to see.

5. Posterior pancreatic veins, tributary to the hepatic portal from the right end of the pancreas.

6. Duodenal vein, the long branch to the hepatic portal which drains the duodenum.

Now trace the hepatic portal posteriorly through the substance of the panceras to a point dorsal to the mesentery of the caecum where the dark sperical *spleen* should be found.

7. Splenic veins. There are several which pass from the spleen into the hepatic portal.

8. Inferior mesenteric vein. It pasess along the dorsal surface of the large intestine receiving blood from numerous small veins in the intestinal wall.

9. Common mesenteric trunk. This is joined near the caecum by the inferior mesenteric. Six more mesenteric veins converge in the mesenteries to form the common trunk which enters the hepatic portal just posterior to the spleen.

E. THE SYSTEMIC VEINS, figs. 30-32.

The tributaries of the four venous trunks noted above (B) will now be studied. These trunks with their tributaries constitute the systemic veins. Since the branches of the precavals are alike on the two sides you need only follow them on one side beginning at the precaval of that side. Those on the left should be traced unless they are better injected on the right. Identify as many as possible of the following veins:

1. Thyroscapular vein, just anterior to the pericardial sac. The precaval receives four tributaries, the most medial of which is the small thyroscapular, which receives a small *thyroid vein* from the gland of that name which is located at the bifurcation of the large arteries just in front of the heart and another, the *scapular vein* from the subscapular muscles in the shoulder region.

2. Internal jugular vein. This joins the precaval just lateral to the thyroscapular and runs along the side of the neck into the head to enter the floor of the skull. It accompanies a large artery (carotid) and a nerve (vagus) in its course alongside the neck. From the esophagus it receives a network (plexus) of small *esophageal veins*. Anteriorly, the internal jugular leaves the base of the skull where it also anastomoses with the external jugular, later to be described.

3. Subclavian vein, the third and largest of the four tributaries to the precaval. It parallels the side of the neck but lies dorsal to the internal jugular.

4. Axillary vein. As the subclavian turns toward the axilla (armpit) it is termed the axillary vein and shortly it proves to be formed of two large branches (5 and 6 which follow).

5. External jugular vein, the branch coming from the head and neck. It anastomoses with the internal jugular where the latter enters the skull. Several *vertebral veins* enter the external jugular at intervals as it parallels the dorsolateral side of the neck. The anterior-most vertebral vein joins it near its union with the axillary and also con-

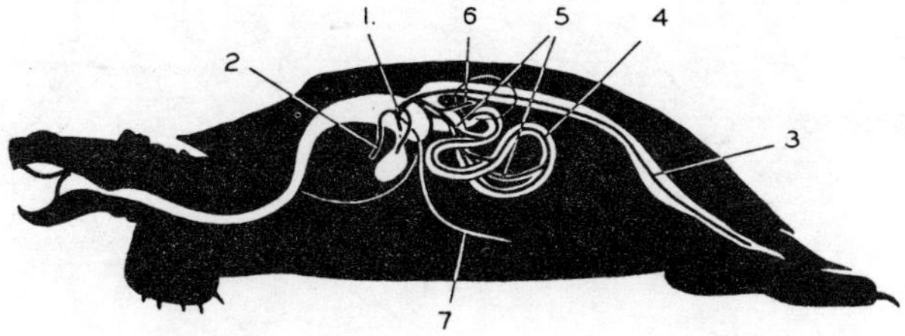

Fig. 33—Hepatic portal syst. Lateral. (Nos. as those in Fig. 34).

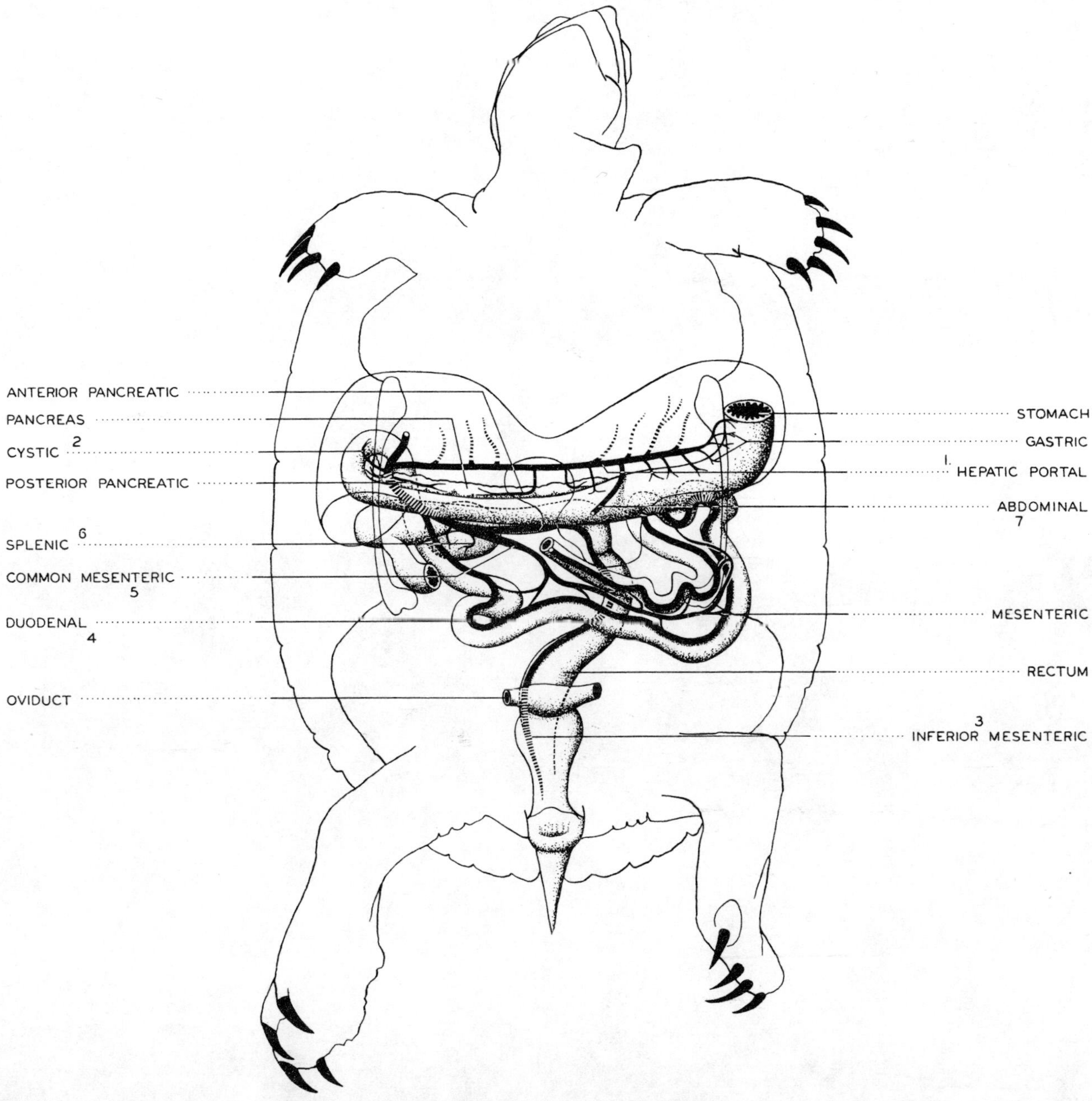

Fig. 34—Hepatic portal syst., ventral

nects with the vertebral vein previously described under the renal portal system as arising near the articulation of the last cervical and the first thoracic vertebrae.

6. Brachial vein. It drains the foreleg and continues into the axillary. The brachial is a little smaller than the external jugular.

7. Scapular vein, the fourth tributary of the precaval. It is the most dorsal and the most lateral of the four tributaries under discussion. It drains the muscles of the scapular region. The left hepatic vein has been described. The postcaval vein was also identified but not its tributaries. These will now be described.

8. Hepatic veins. Many of these enter the postcaval from various parts of the liver. Puncture the postcaval just ouside the sinus venosus and pass your probe within this vein to the liver. Dissect liver tissue to reveal hepatic veins. Now find the postcaval where it enters the liver posteriorly to the right of the hepatic portal vein. Free it from the membranes (serosa of liver and pleuroperitoneum) which bind it between liver and lung and trace it to its bifurcation between the kidneys. These are the roots of the postcaval which are made up of the following veins:

9. Renal veins. These drain the kidneys into the two roots of the postcaval.

10. Genital veins. Several of these also emerge from the kidneys bringing venous blood from the gonads and their ducts.

F. AORTIC ARCHES, figs. 35 and 36, also 26-29.

The conus arteriosus of the turtle looks like three large arteries extending cephalad from the ventricle. Clear away loose tissue so as to be able to separate and identify them from left to right as:

1. Pulmonary artery, the left trunk. It takes venous blood from the right ventricular chamber and, upon forming right and left branches, distributes it to each lung.

2. Left aorta, the middle trunk. It curves dorso-laterad and unites with the right aortic arch posteriorly and in the dorsal midline.

3. Right aorta, the third trunk comprising the conus arteriosus. It is covered ventrally by the large brachiocephalic branch which it gives off just after leaving the ventricle.

4. Brachiocephalic (innominate) artery, mentioned above. It is the main arterial trunk supplying branches to the forelimb, neck, and head. It bifurcates to form two large branches each of which gives off two branches in close succession. Just anterior to the branchings is a round reddish *thyroid gland.* Clear away loose tissue from these vessels.

5. Coronary arteries. The small vessels which may be seen arising from the brachiocephalic and coursing over the surface of the heart.

6. Subclavian artery. The right and left subclavians are the large branches formed by the bifurcation of the brachiocephalics.

7. Thyroid arteries. These arise from the subclavians near the origins of the latter and nourish the circular thyroid gland.

8. Ventral cervical artery. A given subclavian sends several small arteries to the ventral surface of the neck. The largest of these is the ventral cervical artery. Ventral cervicals also supply branches to the neck muscles, trachea, esophagus, and thymus glands — yellowish tissue masses lateral to the ventral cervical arteries.

9. Axillary artery, the continuation of the subclavian where it loops sharply dorsal from a point ventral to the scapula. It usually trifurcates to form the three following arteries:

10. Brachial artery, the continuation of the axillary. It arises where the axillary bends laterally on itself.

11. Vertebral artery, a large branch of the axillarv which runs dorso-mesad to send branches between the dorsal ends of the fused ribs as the *intercostal arteries* which parallel the sutures of the costal plates and anastomose laterally with the *marginocostal artery.* Vertebral arteries may extend anteriorly from the axillary, as well as posteriorly.

12. Dorsal cervical artery. This arises from the axillary near its junction with the brachial, or it

may be a branch of the vertebral at a point roughly ventral to the anterior margin of the carapace. The dorsal cervical obviously supplies muscles, skin, and other tissues on the dorsal side of the neck.

13. Common carotid artery. This passes cephalad along the ventral side of the neck (it may cross the subclavian). The carotid may form a loop in the neck if the latter is withdrawn between carapace and plastron. Notice small branches to the thymus gland from the carotid. Anterior to this point it continues cephalad in company with the vagus nerve and the internal jugular vein. The carotid enters the head via a foramen located just anterior to the auditory area of the skull.

G. AORTIC ARCHES AND THEIR BRANCHES, figs. 35 and 36.

The right and left aortic arches each curve laterally and dorsally as they leave the ventricle. Trace the left arch, separating the left lung from the stomach (if this was not previously done) and turning the stomach ventrally and to the right. The left arch passes to the left of the esophagus and dorsal to the stomach and liver where it forms three large branches.

1. Gastric artery. This supplies the cardiac region of the stomach where it bifurcates forming *ventral* and *dorsal gastric arteries,* the former supplying the lesser curvature and the latter supplying the greater curvature of the stomach.

2. Coeliac artery. It branches an inch or less from the left aortic arch to form the following two arteries:

3. Anterior pancreaticoduodenal (to the left end of the pancreas). It supplies the pyloric end of the stomach and part of the liver and branches to the right, sending numerous small arteries into the pancreas, duodenum, and liver.

4. Posterior pancreaticoduodenal, the second branch of the coeliac. It supplies the right end of the pancreas, duodenum, liver, and, as the *cystic artery* the gall bladder.

5. Anterior mesenteric artery, the third branch of the left aorta. This passes posteriorly in the center of the intestinal mesenteries where it subdivides into numerous branches which radiate through the mesentery to all parts of the small intestine.

6. Posterior mesenteric artery, one of the branches of the superior mesenteric which continues to the large intestine which it accompanies all the way to the cloaca where it sometimes anastomoses with one of the common iliac arteries.

7. Junction of right and left aortae. This can be seen an inch or less posterior to the point at which the anterior mesenteric artery originates from the left aorta. To facilitate tracing the aorta the postcaval vein may be removed. Trace the right aorta to its origin in the ventricle of the heart. The only branch from the right aorta is the brachiocephalic, noted above.

8. Dorsal aorta, the unpaired posterior part of the aorta which continues caudad in the dorsal midline just ventral to some prominent longitudinal bands of muscle to which small arteries are supplied from the aorta.

9. Genital arteries (*ovarian* in females and *spermatic* in males). One or more arises from the aorta to supply arterial blood to each gonad.

10. Renal arteries (two or three to each kidney as the aorta passes between them). These are rather small and are revealed by removing connective tissue along the aorta between the kidneys.

11. Epigastric artery. This leaves the aorta just dorsal to the kidney which should be separated from the carapace and together with the gonad turned toward the opposite side. Follow the epigastric artery laterally to the place where the renal portal vein enters the pleuroperitoneal cavity. At this point the epigastric branches, sending an artery anteriorly along the border of the carapace to join the marginocostal artery which also follows the margin of the carapace.

12. Common iliac artery, one of two branches of the dorsal aorta just posterior to the origin of the epigastric. Each common iliac promptly divides into an internal and an external iliac.

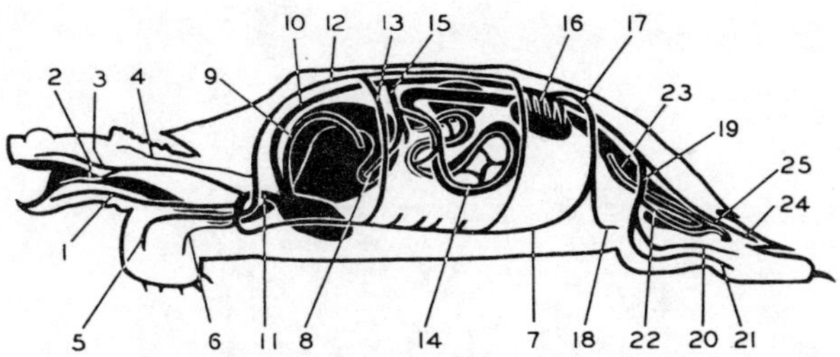

Fig. 35—Arteries, lateral view (nos. as those in Fig. 36).

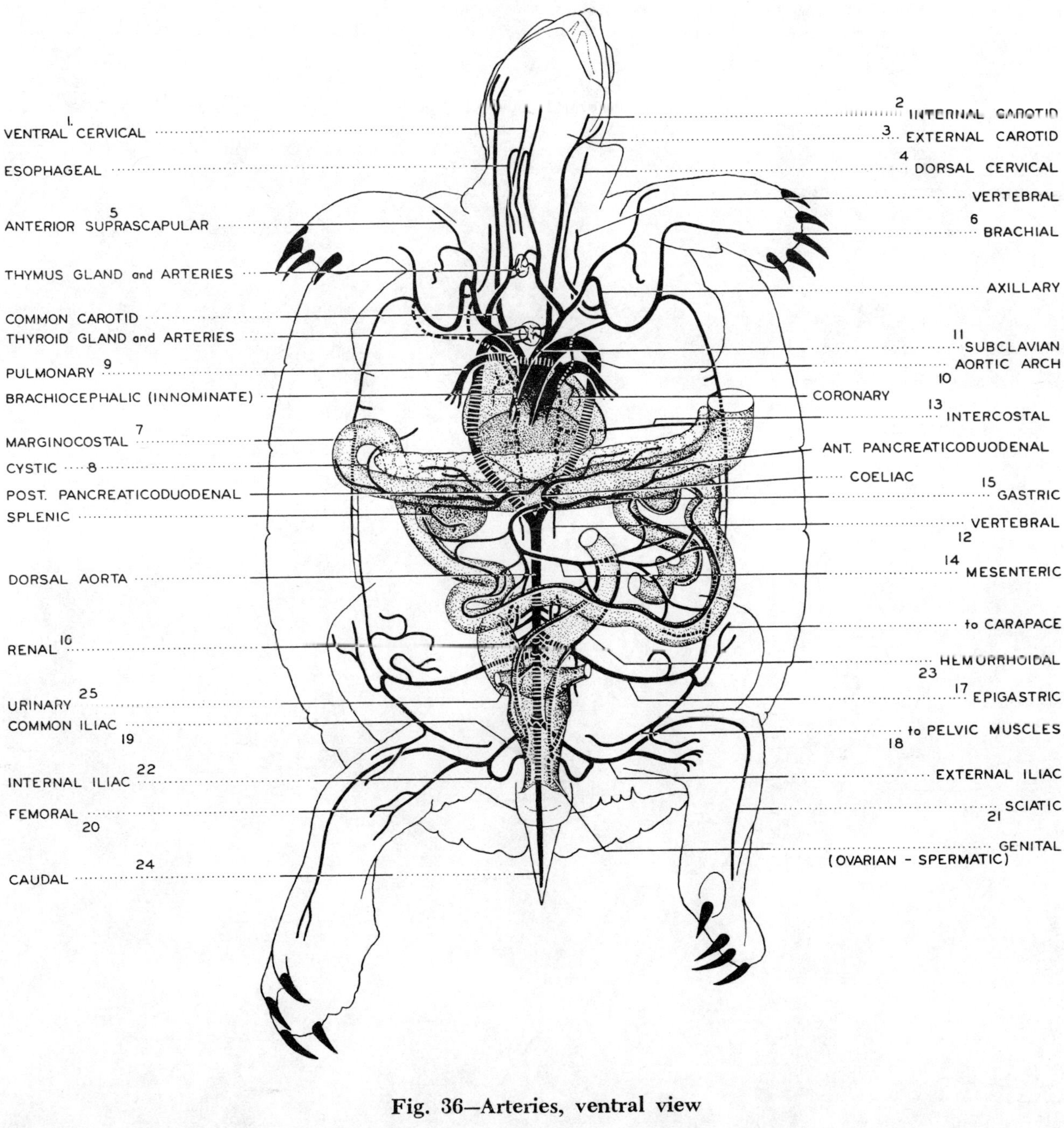

Fig. 36—Arteries, ventral view

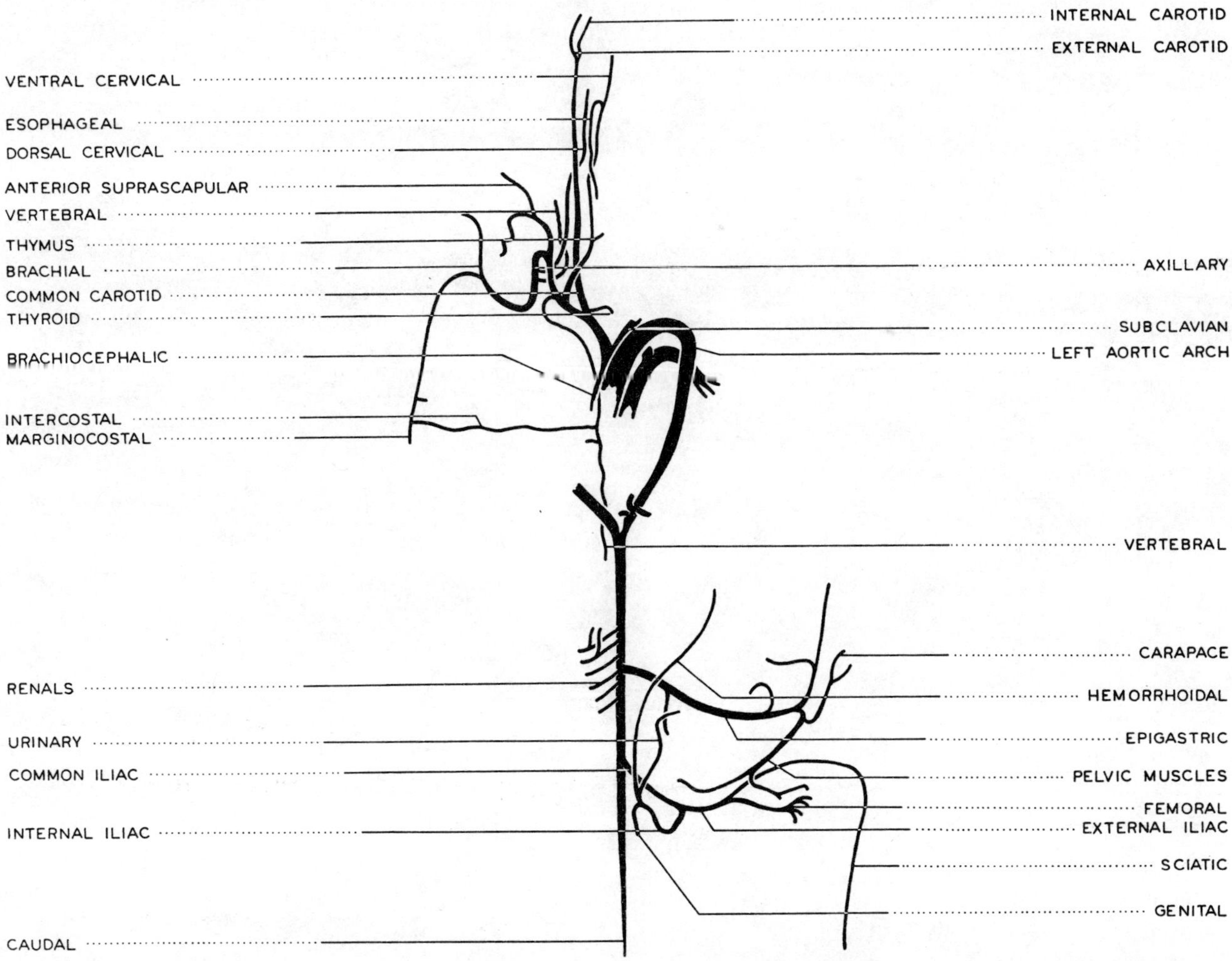

Fig. 36A—Variation from typical arterial system seen in Fig. 36.

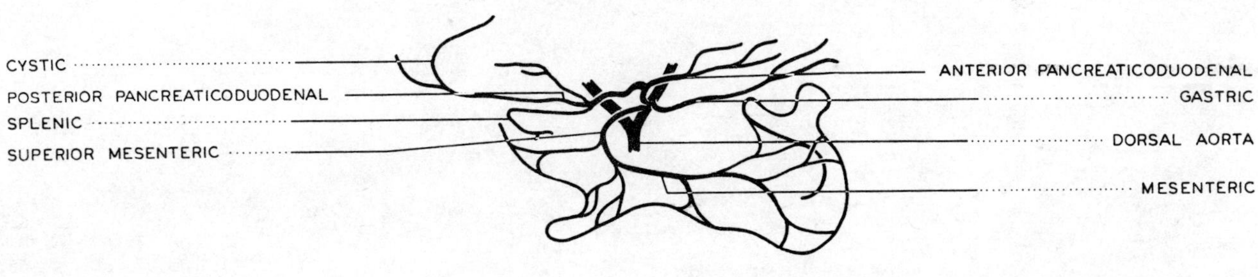

Fig. 36B—Variation of visceral arteries seen in Fig. 36.

13. Internal iliac. It sends branches to the reproductive organs, pelvic region and bladder, but its largest branch is the:

14. Haemorrhoidal artery. This supplies the large intestine, passing cephalad along its side. It may be larger on one side than on the other, one giving more blood to the large intestine while the other, being smaller, permits additional blood to enter the pelvic organs via the internal iliac.

15. External iliac, a vessel that divides soon after leaving the internal iliac. It sends a larger ventral branch through the pelvic region giving off small arteries to the pelvic musculature.

16. Femoral artery, the continuation of the external iliac into the pelvic area and to the thigh.

17. Sciatic artery, one which passes dorsad from the external iliac to the sacroiliac joint, traverses a nerve dorsally and turns ventrad into the thigh following the medial side of the ilium as it leaves the pelvic area. *This is the principle artery of the leg.* Bone cutters will be useful in tracing this artery as well as in tracing the femoral.

18. Middle sacral or caudal artery, a continuation of the dorsal aorta caudal to the origins of the common iliacs. It supplies the dorsal side of the cloaca and the tail region.

CHAPTER VI

Urogenital System

Remove all but the posterior-most one inch of the large intenstine, all of the small intestine and the stomach. (If you desire to retain these organs simply reflect them anteriorly after cutting the large intestine an inch anterior to the anus and severing the mesenteries along the intestinal tract.)

A. FEMALE, figs. 37 and 38.

1. Ovaries, previously noted as large egg-filled sacs on each side in the posterior pleuro-peritoneal cavity. Note the yellow eggs in various stages of development (eggs may all be very small in young females).

2. Mesovarium, the mesentery of the ovary. This vascular membrane supports the ovary by attachment to the kidney peritoneum.

3. Oviduct (Mullerian duct), a broad white coiled tube lying along the posterior side of each ovary. Adjacent to the ovary its cephalic end opens as the *ostium*. Carefully search for the ostium where the mesovarium splits near the anterior end of the oviduct. Compare with amphibian and elasmobranch. The posterior end of the oviduct opens into the antero-lateral wall of the cloaca, ventral to the rectal opening.

4. Urinary bladder, a large bilobed sac narrowing posteriorly where it empties into the cloaca between the two oviducts. Open the bladder ventrally and observe the texture of its inner surface, then probe its neck posteriorly until your probe enters the cloaca ventral to the oviducal openings. NOTE: If not previously opened the cloaca should now be slit postero-anteriorly and a little to the left of its mid-ventral line, but first the midregion of the pelvic girdle must be removed with bone shears. Avoid damage to the cloaca which is generally blackish along its ventral side.

5. Cloaca, a cylindrical chamber opening via the anus posteriorly and receiving the contents of the oviducts, bladder, and large intestine or rectum.

6. Accessory urinary bladders, thin sacs on either side of the cloaca to which they are attached laterally and at a point posterior to the oviducts. They are known to carry water in females for use in softening the earth while nest-digging. To find these you should first clear away the connective tissue along-side the cloaca. Slit and probe one of these bladders so as to see its opening into the cloaca.

7. Clitoris, a series of dark thickenings in the ventral wall of the cloaca. This mass is homologous to the male penis.

8. Kidneys (metanephroi), previously identified as lobed structures closely flattened against the postero-dorsal wall of the pleuroperitoneal cavity on either side. On the ventral side of each kidney may be seen the renal portal vein and its tributary, the internal iliac vein.

9. Ureter (metanephric duct), a tube which lies just dorsal to the renal portal vein. This vein should be carefully removed over the ventral face of the kidney so as to disclose the ureter. Make a small slit in the ureter opposite the kidney and pass a slender probe through it to its cloacal orifice which lies just anterior to that of the oviduct.

B. MALE, figs. 39 and 40.

Expose the cloaca, if not previously done, as directed for the female urogental system, A-4 above, but open it by slitting it up one side to avoid injury to the penis. Observe the opening of the rectum dorsally and that of the urinary bladder ventrally into the cloaca at its anterior end.

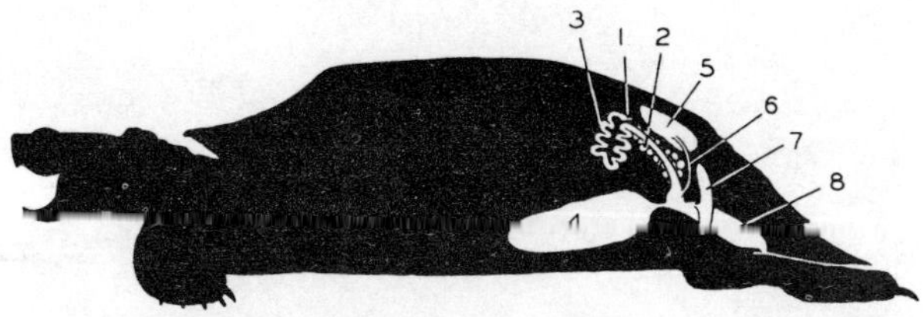

Fig. 37—Female urogenital syst. lateral (nos. as those in 38).

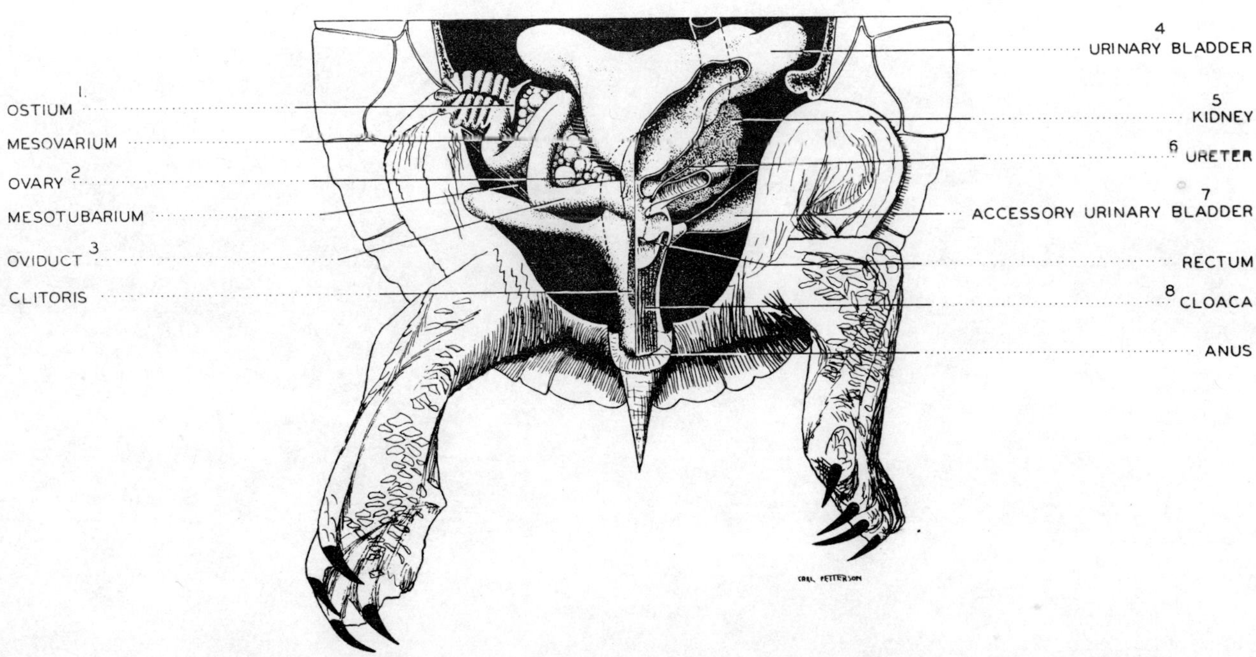

Fig. 38—Female urogenital syst., ventral

1. Penis, a dark mass visible through the ventral cloacal wall. A penis is not present in anamniotes but does occur in reptiles, also in some birds, and in all mammals. Attached to the ventral cloacal wall are muscles which retract the penis. Its eversion is achieved by vascular engorgment of the paired *corpora cavernosa penis,* rounded spongy tissue masses projecting laterally at each side of the neck of the bladder from the anterior cloacal wall. Its midlongitudinal *urethral groove* terminates caudad in a heart-shaped *glans penis.*

2. Kidneys, previously seen as lobed organs flattened against the postero-dorsal walls of the pleuroperitoneal cavity.

3. Testes, male gonads. Each testis is a spherical organ of yellowish color attached to the ventral side of each kidney by a mesentery termed the *mesorchium.*

4. Epididymis, a dark, elongated structure of coiled tubules, one of which lies postero-lateral to each kidney. Each testis receives, in its anterior pole, a series of minute *efferent ductules* from the testis of its own side. These ductules convey sperm cells from testis to epididymis and they traverse the mesorchium en route. Efferent ductules and epididymis are derived from embryonic mesonephric tubules and ducts.

5. Ductus deferens (deferent ductule), derived from the mesonephric duct also. It is much coiled anteriorly where it may be termed the epididymidal duct. Remove its peritoneal covering and uncoil it posteriorly, thus tracing it to its cloacal orifice which lies cephalad of, and basal to, the bulbs of the corpora cavernosa. These bulbs, which when the penis is retracted extend craniad along either side of the bladder-stalk, are the anterior swellings of the corpora cavernosa. Filling of the corpora cavernosa with blood from the internal iliac vein causes erection of the penis at the time of mating. The walls of the urethral groove meet dorsally to form a duct for the passage of spermatozoa.

6. Ureters. These convey urine from kidneys to the cloaca. The ureters lie on the kidney's ventral face and may be seen by removing the epididymis from the kidney. Each ureter is a short uncoiled tube the cloacal orifice of which is just anterior to that of the ductus deferens of the same side. These openings are at the anterior end of the urethral groove.

7. Accessory bladders. As in the female, A-6 above. Find and probe these bladders to discover their openings into the cloaca.

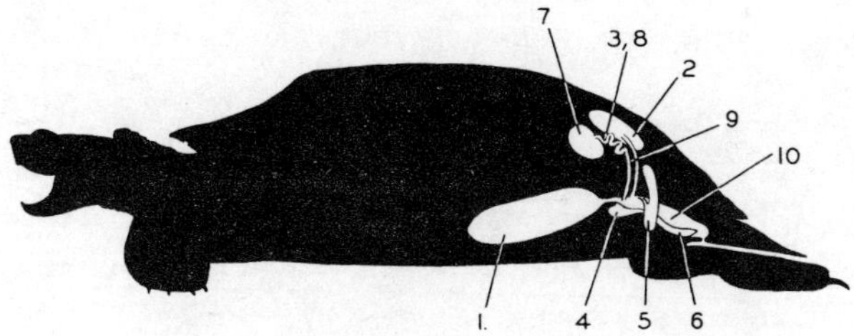

Fig. 39—Male urogenital syst. lateral (nos. as those in 40).

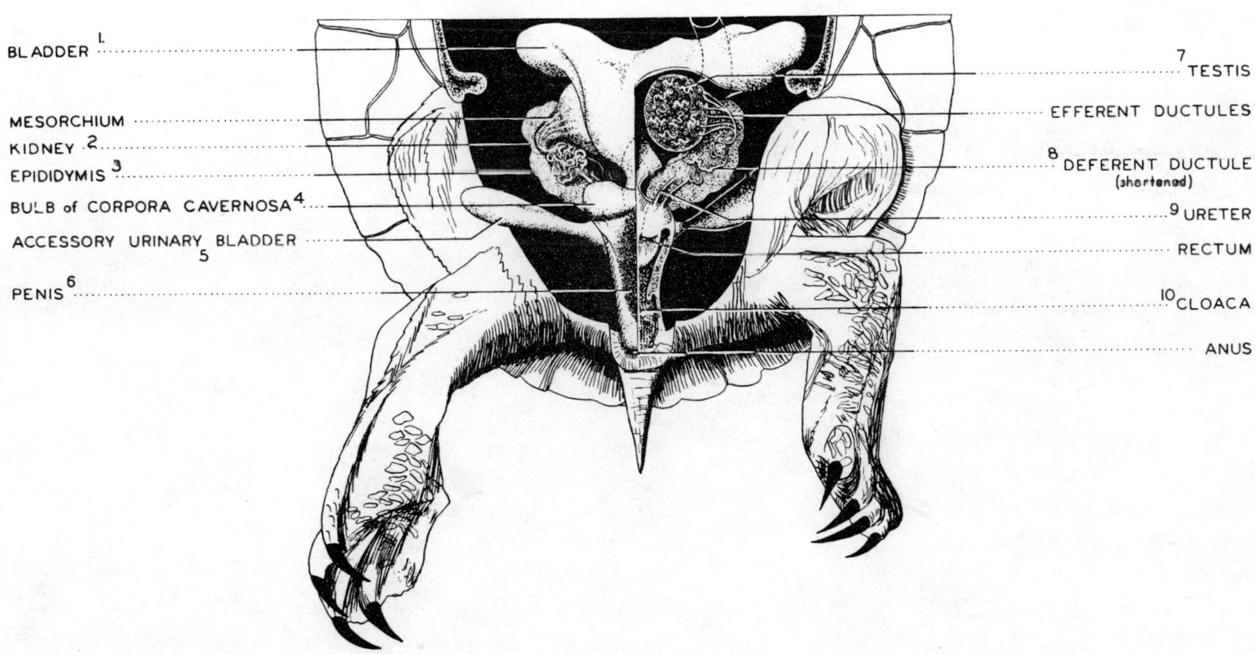

Fig. 40—Male urogenital syst. ventral.

CHAPTER VII

Nervous System

The brain and spinal cord are relatively smaller in the turtle than in animals with cartilaginous skeletons such as the shark. However, a careful study of the nervous system of the turtle is both interesting and profitable owing to its higher degree of development and complexity. Care must be exercised in opening the skull as well as in removing the brain lest important structures be damaged or destroyed. A pair of bone shears will be useful in removing the top of the skull (skull cap). This permits study of the dorsal view of the brain without its being disturbed. This view should, therefore, be studied before the ventral and lateral surfaces are examined.

THE BRAIN

A. DORSAL SURFACE, fig. 43.

This is clearly seen only after complete removal of the skull "cap," as directed above, and the meninges now to be described.

1. Meninges, membranes covering the brain and spinal cord.

a. Dura mater, a tough membrane lying between the brain and the skull.

b. Pia mater, a thin membrane which intimately covers the brain surface to which it is closely adherent.

c. Subdural space, a narrow space lying between the two above membranes. It is crisscrossed by numerous membranous cords passing between the dura and the pia maters.

d. Peridural space, a narrow space lying between the skull and the dura mater.

Remove as much as you can of the dura mater from the dorsum of the brain.

2. Cerebral hemispheres, a pair of conspicuous swellings forming two halves of the cerebrum, the largest of the several regions of the brain. The hemispheres are separated midlongitudinally by a deep *superior longitudinal fissure.* In turtles they are relatively much larger than those of fishes and amphibians. The lateral surface of each hemisphere is extended to form a flattened *temporal lobe.*

a. Olfactory bulbs or lobes, a pair of elongated triangular anterior extensions from the cerebral hemispheres. These bulbs each receive an *olfactory nerve* from an *olfactory sac,* (one sac on either side in the nasal cavities).

b. Choroidal sac, a very thin sac adhering to the dura mater. It is often torn off when removing the dura. This "sac" lies between the posterior ends of the cerebral hemispheres where it forms a *choroidal roof* for the *diencephalon.*

c. Pineal body (epiphysis), a small body lying upon the choroidal sac to which it is attached. It lies in the triangle formed by the posterior separation of the cerebral hemispheres.

Remove the choroidal sac and note the cavity of the diencephalon which is a shallow depression in the brain floor just caudad of the cerebral hemispheres.

3. Optic lobes, a pair of prominent subspherical bodies just caudad of the cavity of the diencephalon. They form a large part of the midbrain region being important centers in the visual system.

4. Cerebellum, a posteriorly projecting pair of lobes which overhang the anterior end of the medulla oblongata. The cerebellum is relatively smaller than that of selachians (elasmobranchs)

but larger than that of amphibians. It functions as the principal center of equilibrium and muscular coordination of the brain.

5. Medulla oblongata (hindbrain), a caudally tapering region continuous with the spinal cord posteriorly.

a. Tela choridea, a thin roof of the fourth ventricle or cavity of the medulla. The tela choroidea is similar in structure and function to the choroid roof of the diencephalon, each serves to secrete nutritive cerebral fluid into the cavities (ventricles) of the brain.

b. Somatic sensory columns, the dorsal margins of the medulla.

c. Auricular lobes, small lobations of the cerebellum representing the anterior continuations of the somatic sensory columns. These columns convey sensations to the cerebrum and cerebellum from outside the body and from the body wall.

d. Somatic motor columns, fiber bundles paralleling the midventral groove in the floor of the medulla and conveying impulses for muscular contraction from the cerebrum and cerebellum to the somatic muscles.

6. Cranial nerves, figs. 41-43. In the turtle, as in amniotes in general, there are twelve pairs of cranial nerves instead of but ten pairs as in anamniotes. You should memorize both the number and the name for each of the cranial nerves.

I. Olfactory nerves. There is one from each olfactory sac to each olfactory bulb. They carry impulses for the sense of smell.

II. Optic nerves. These will be seen after cutting through the olfactory nerves and lifting the anterior end of the brain slightly. Optic nerves are rather stout trunks which carry visual impulses from the eyes to the optic lobes. These nerves enter the cerebrum ventrally on either side.

III. Oculomotor nerves. They leave the floor of the midbrain just anterior to the trochlear nerves (IV). They are visible only after pressing the cerebral hemisphere away from the skull. The oculomotor nerve can be further traced by loosening the eyeball ventrally and lifting it so as to reveal the loose tissues between the eyeball and the orbital floor. Remove loose connective tissue down to the glandular tissues so as to be able to see these nerves. Usually adherent to the eyeball is a portion of the maxillary branch of the trigeminal (V). Loosen this nerve from the eyeball, cut the pyramidal muscle, lift the eye upwards and mediad, then cut the inferior and external rectus muscles between and above which will be found the heavy white trunk of the optic nerve (II). The oculomotor nerve contacts the ventral side of the optic nerve and supplies the four following extraocular muscles: superior, inferior, and anterior rectus, and inferior oblique.

IV. Trochlear nerves. On the dorsal side of the midbrain, in the dorsolateral angles between optic lobe and cerebellum, will be seen the small trochlear nerves. By pressing the cerebral hemispheres away from one orbit a trochlear nerve will be seen to pass behind the oculomotor toward the orbit which it enters.

Cut the superior oblique muscle at its insertion on the eyeball and look beneath this muscle to find the trochlear nerve which innervates it. Another nerve running medial to the trochlear is the *ophthalmic branch* of the trigeminal (V).

V. Trigeminal, the thickest of the cranial nerves. These originate from about the middle of the ventral wall of the medulla and pass laterad to enter a large semilunar ganglion located in a fossa in the medial wall of the skull. From the ganglion, on either side, three branches pass anteriorly one above another. From above they are the *ophthalmic* distributed to tissues around the eyes, the *maxillary* innervating the face, and the *mandibular* distributed to the lower jaw. Uncover the ophthalmic and maxillary nerves by removing an eyeball without injury to these nerves. Sever the nerves lying anterior to the trigeminal and raise the brain gently drawing it toward one side to reveal the ophthalmic nerve. This nerve may now be traced anteriorly into the olfactory sac. Traced from the orbit to the brain the ophthalmic enters the skull, travels in company with the troch-

lear nerve between the dura mater and the skull and enters the semilunar ganglion from which its fibers, as part of the main trigeminal trunk, pass into the brain. The maxillary nerve traced posteriorly comes from two branches, one passing below the orbit and the other coming up diagonally across the floor of the orbit to join its mate at the posterior orbital wall. From here the maxillary passes between muscles to enter the skull. At this point it is joined by the mandibular branch coming up from the lower jaw. The trunk formed by the union of these two nerves enters the skull through a foramen and promptly joins the semilunar ganglion from which its fibers parallel those of the ophthalmic as they pass into the brain.

VI. Abducens nerves, these spring from near the midventral line medial to the roots of the facial (VII) and acoustic (VIII) nerves. Do not attempt to see the abducens until after the remainder of the cranial nerves have been traced.

VII. Facial nerves. These arise in company with the acoustic (VIII) nerves from the lateral wall of the medulla and just posterior to the trigeminal roots. They at once separate into the facial nerve passing through the capsule of the inner ear anteriorly and the acoustic next to be described. The facial soon passes rostrad out of the skull over the anterior horn of the hyoid bone and innervates muscles between the hyoid and the lower jaw. In order to further trace the seventh as well as the ninth, tenth, and twelfth nerves proceed as follows: remove the skin and adjacent muscles from the hyoid bones on the ventral side of the head, locate the anterior and posterior horns of the hyoid, and identify the nerves mentioned above, in turn, as described below.

VIII. Acoustic (auditory) nerve. It arises in company with the facial and passes promptly into the inner ear capsule to divide into an acoustic branch to the cochlea and a vestibular branch to the semicircular canals and vestibule.

IX. Glossopharyngeal nerve. It arises as several small roots just posterior to the acousticofacial roots on the lateral walls of the medulla. These nerves exit from the skull by passing through the posterior wall of the ear capsules and forward between the two horns of the hyoid on either side to supply adjacent muscles as well as the lining of the mouth and pharynx.

X. Vagus nerve. It springs from several rootlets along the side of the medulla just posterior to the ninth nerve roots. These two nerves leave the skull together with the eleventh and twelfth nerves. Now locate the vagus (vagosympathetic) trunk in the neck and follow it into the skull by dissecting longitudinally through the floor of the mouth and pharynx into their cavities. Turn back the two flaps formed by this incision and remove the mucous membrane from the roof of the pharyngeal cavity. Here the hypoglossal nerve (twelfth) will be seen passing ventral to the vagus. At about this point the vagus enters the *superior cervical sympathetic ganglion* from which several nerves emerge, some (or one) of which enter the skull and medulla as the vagus roots.

XI. Spinal accessory. Like the vagus the spinal accessary is formed from several roots which emerge from the posterior medulla to exit from the skull in company with the vagus. The accessary nerves supply motor impulses to the muscles of the neck.

XII. Hypoglossal nerves. These emerge from the posterior medulla near the mid-ventral sulcus. They can best be seen by first cutting the roots of the vagi and accessories on one side so as to uncover them. Traced rostrad into the throat the hypoglossal supplies motor fibers to muscles of the anterior hyoid horn and to the tongue.

B. Ventral view of the brain, fig. 41. Removal of the brain from the skull is required in order to study the ventral (basilar) surface. First, gently lift the olfactory bulbs and (if not done before) cut the optic nerves which lie side by side just beneath the bulbs.

1. Optic chiasma, a fusion of the two optic nerve trunks beneath the olfactory bulbs. A partial crossing of nerve fibers to the opposite side occurs at this point.

2. Infundibulum, a median thin saccular depression of the diencephalic floor just posterior to the optic chiasma. Attached by a slender thread-like stalk is the ventrally placed:

3. Pituitary body (hypophysis), a yellowish endocrine gland about half the size of the infundibulum. It is firmly seated in a bony cavity, the homologue of the mammalian *sella turcica* (Turk's saddle). Over the pituitary body is a thin membrane, the *diaphragma sellae* which should be torn with a teasing needle passed around the hypophysis. Now the hypophysis may be loosened underneath and lifted out. It is difficult to avoid breaking the infundibular stalk in this dissection.

C. Lateral view of the brain, fig. 42.

Study the lateral surface of the brain and identify each of the parts that were identified on the dorsal surface. Notice exactly from what part of the brain each of the twelve pairs of cranial nerves arises. Determine the relationships between pituitary body, infundibulum, and diaphragma sellae. Can you identify the *mammilary body* which is a slight bulge from the posterior wall of the infundibulum just dorsal to the pituitary body? Notice also the temporal lobes, optic lobes, and the cerebellum.

D. Median sagittal section, fig. 44.

Section the brain midsagittally with one clean stroke of a long-bladed knife. This should give a smooth-cut surface which will show details better than would a ragged surface. Study the internal structures of the brain. Cut open a hemisphere and observe its cavity or *ventricle*. (The first two ventricles are in the two hemispheres, the third is largely in the diencephalon and the fourth is partly beneath the cerebellum and partly within the medulla oblongata). In each case, compare the size of the ventricle with the thickness of the

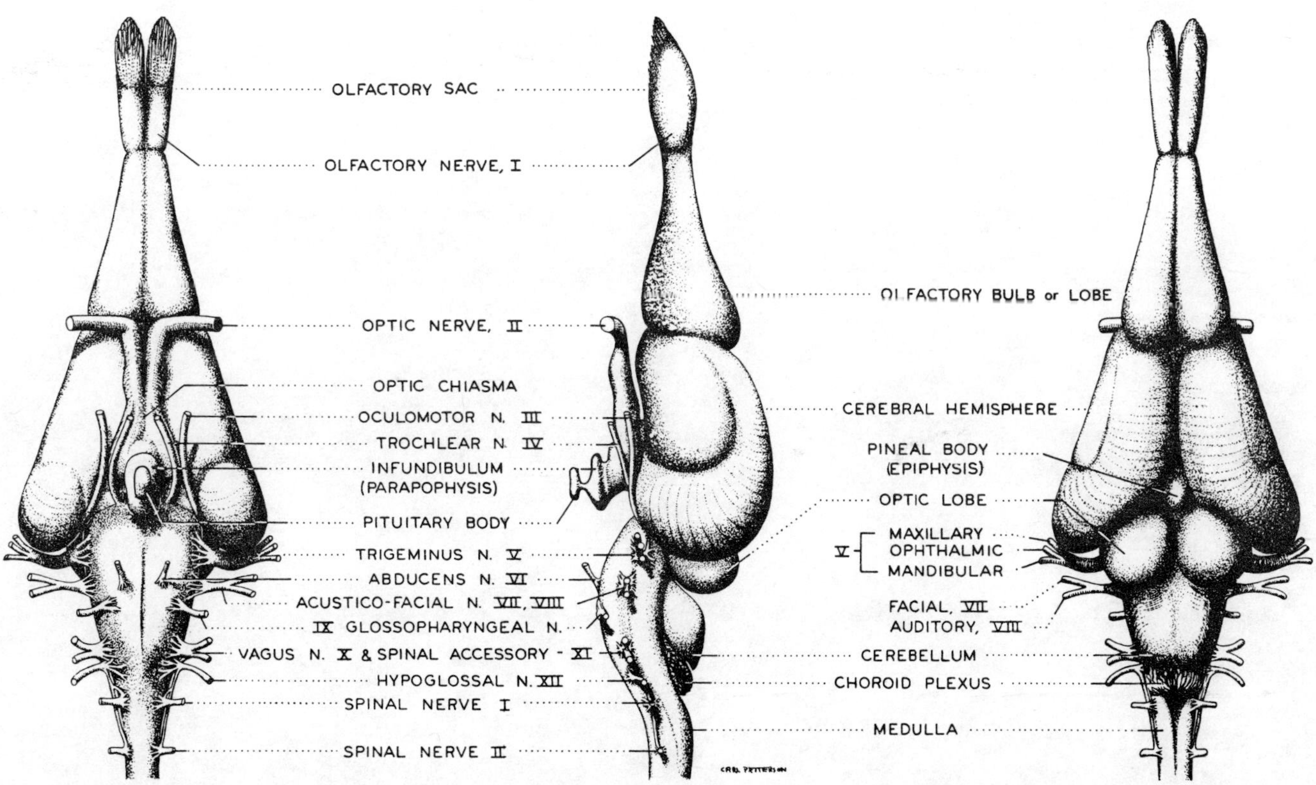

Fig. 41—Brain, ventral **Fig. 42—Brain, lateral** **Fig. 43—Brain, dorsal**

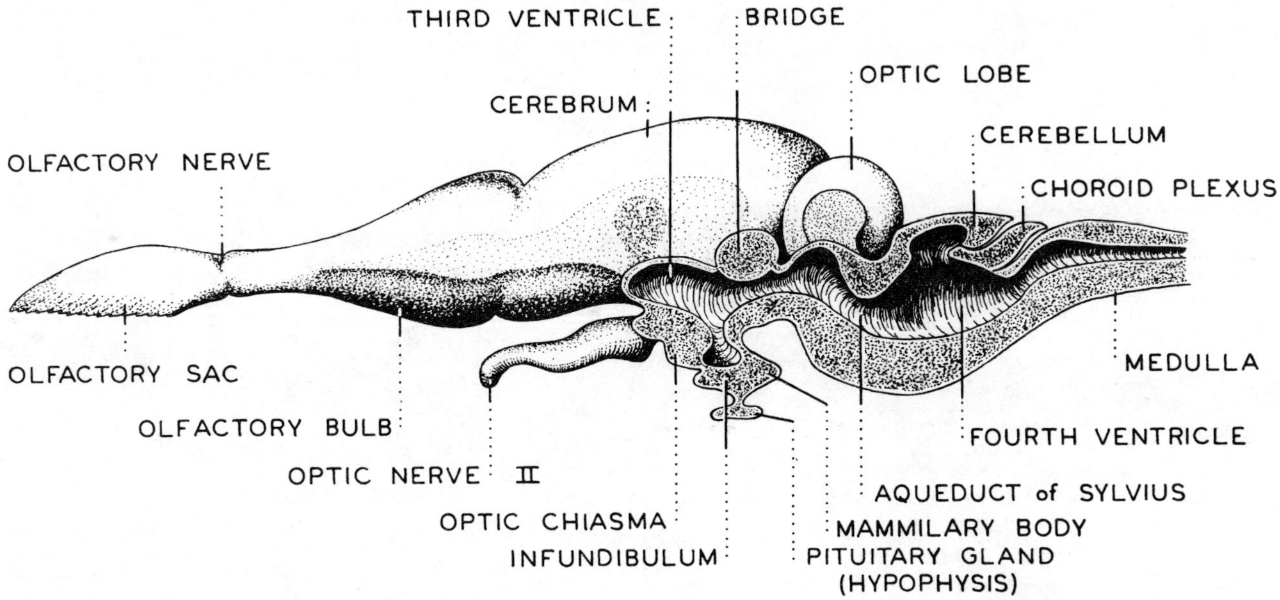

Fig. 44—Brain, midsagittal

wall of that part of the brain containing it. Try and find a communication between the first two ventricles and the *third ventricle*. These small openings are the *foramina of Monro*. Observe the *cerebral aqueduct* (aqueduct of Sylvius) which connects from the posterior end of the third ventricle to the anterior end of the *fourth ventricle*. The aqueduct traverses the length of the *midbrain* which lies between the inferior portions of the optic lobes. Further open and observe the cavity of the optic lobe and note the cavity of the cerebellum just posterior to it. The cavity of the fourth ventricle continues, much reduced in size, down the center of the spinal cord as its *central canal*. This canal is so small as to be almost microscopic in size.

Additional features of the midsagittal section are:

1. The pallium (roof) of the cerebral hemispheres.

2. Corpus striatum, a large mass projecting up from the floor of each lateral ventricle. This mass contains groups of brain cells (nuclei or "ganglia") of the cerebrum.

3. Epithalamus, the entire roof of the diencephalon. This roof is largely overlain by the cerebral hemispheres.

4. Thalamus, a bulge from each lateral wall of the diencephalon. It includes important neurones for relaying sensations from the body to the cerebrum.

5. Hypothalamus, the ventral part of the diencephalon. It includes the infundibulum, hypophysis, and mamillary bodies.

6. Optic ventricles, the cavities of the optic lobes. These open into the third ventricle.

7. Choroid plexus, a delicate mass of tissue extending from the wall of each of the ventricles. Note carefully the relations between choroid plexus and the ventricle wall in each case. Are these relations identical?

THE SPINAL NERVES

These nerves are attached to each side of the spinal cord by sensory (dorsal) roots and by motor (ventral) roots, fig. 45. There are typically nine pairs of cervical spinal nerves in the neck and ten pairs of thoracic spinal nerves in the trunk. In general a pair of these nerves is developed for each vertebra in the series.

1. Brachial plexus, the fibers of the sixth, seventh, and eighth cervicals plus a small branch from the first thoracic. These nerves anastomose freely in the region of the axilla to form the brachial plexus of nerves. This plexus supplies three main nerves to the foreleg, namely the dorsal radial, the ulnar and the median.

2. Lumbosacral plexus, the eighth, ninth, and tenth, thoracic nerves together with the two sacral nerves. These anastomose to form the lumbosacral plexus which innervates the posterior appendage on either side. This plexus is located medial to the ilium and from it fibers of the eighth and ninth nerves scatter out to muscles and skin along the outside of the thigh. Fibers from the tenth and two sacral nerves form three sciatic nerves passing among the posterior thigh muscles to the leg and foot.

AUTONOMIC NERVOUS SYSTEM

The autonomic nervous system mediates visceromotor and viscerosensory impulses to the spinal cord, via communicating rami, and brain but is devoid of all voluntary control – it is therefore, in a sense, an automatic system, fig. 45.

1. Anterior head trunk, a group of nerve fibers arising from the trigeminal (fifth) cranial nerve. These fibers anastomose to form the sphenoid plexus as they progress toward the intra-orbital nerve. They continue caudad as a slender trunk which parallels the long axis of the skull. This nerve turns around the ear (otic capsule) which it leaves via an aperture just above the lower jaw articulation. It next extends below the glossopharyngeal nerve and passes between the carotid artery and the vagus and hypoglossal nerves at which point it receives a small trunk formed by branches from both the facial (seventh) and glossopharyngeal (ninth) nerves.

2. Superior head trunk. This is formed by a group of fibers which enter the sympathetic trunk from the glossopharyngeal nerve. This trunk continues caudad receiving additional sympathetic fibers from the vagus, accessory, and hypoglossal nerves.

3. Posterior cervical trunk, a more caudal bundle which receives its fibers largely from the vagus and first cervical nerves.

4. Middle cervical trunk. This is found at the point where the sympathetic trunk branches from the vagus. Posterior to this point the sympathetic trunk passes dorsal to the brachial plexus.

5. Inferior cervical ganglion, a spindle-shaped ganglion (enlargement) a little distance caudad of the point of origin of the middle cervical trunk.

6. Second inferior ganglion, another ganglion on the sympathetic trunk. It lies a little caudad of the first inferior cervical ganglion and from the second ganglion there arise several fine nerves. Caudal to this lies the *first spinal ganglion* close to which, and on the sympathetic trunk, lies the next sympathetic ganglion. Between these two ganglia are strands of communicating fibers (rami communicantes). Posterior to this the main trunk forms one or two ganglia close to each spinal ganglion and sends at least one communicating branch (ramus communicans) to the latter. To simplify tracing these nerves you may remove entirely the lung on the side which you are following.

SPECIAL SENSE ORGANS

A. THE NOSE.

Contained within the nose are the olfactory sense organs. Within the anterior nares are the nasal sacs (previously described) which contain the olfactory mucous membranes which are sensory for smell. A *median septum* separates the right and left nasal passages and olfactory sacs. This septum is stiffened by means of a thin bony plate covered by softer tissues. A *nasal concha* (turbinal) projects from the posterior wall of each nasal cavity. The concha is homologous with the

mammalian nasal concha or turbinate bone. A prominent unnamed fold arises from the floor of each nasal cavity.

The olfactory mucous membrane contains special "neuroepithelial cells" capable of receiving smell sensations. The olfactory nerves (previously described) transmit smell impulses to the olfactory bulbs of the cerebrum.

B. THE EYE.

Turtle eyes are small but if care is exercised they can be dissected and will provide good prac-

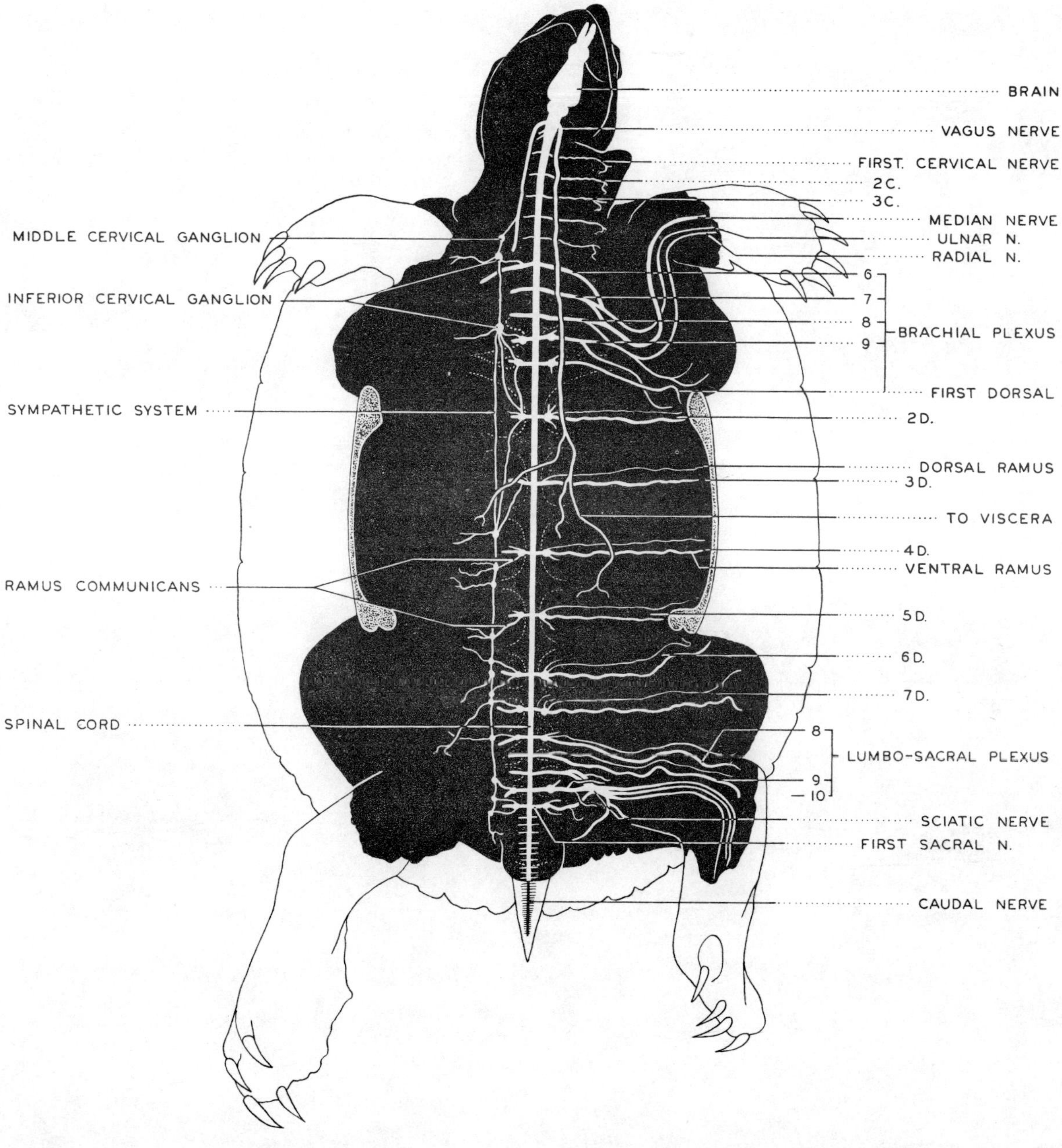

Fig. 45—Spinal nerves and autonomic syst., ventral

tice in minute dissection — of great value to a surgeon. The anatomy is very similar to that of the shark eye. Cut through the skin around the eye and remove the bone between the two eyes with bone shears. Identify the following structures:

1. Interorbital septum, a bony partition between the two orbits. An artery runs along either side of this septum.

2. Harderian gland, a small gland whose function is still disputed. One of these lies on the antero-dorsal side of each eyeball.

3. Lacrimal gland, the tear gland. It is larger than the Harderian gland and it lies against the ventral surface of the eyeball. Remove the lacrimal gland so as to expose more fully the eye muscles and the eyeball.

4. Pyramidalis muscle, a flat muscle covering the antero-ventral surface of the eye.

5. Extra-ocular muscles. These lie outside the eyeball and their contractions control the movements of the eyeball. (*Intra-ocular muscles* also exist for the purpose of adjusting the focus of the eye by changing the lens curvature). Looking directly at the pupil of the eye the extra-ocular muscles will be seen to insert on the eyeball as follows: the *superior rectus* inserts dorsally, the *inferior rectus,* ventrally, the *medial* (*anterior*) *rectus* posteriorly. These four rectus muscles take their origin together near the center of the depth of the orbit. The *superior oblique* muscle extends from the interorbital septum to the dorsum of the eye while the *inferior oblique* originates on the interorbital septum and inserts on the ventral surface of the eye. Observe the trochlear nerve passing over the medial rectus on its way to the superior oblique and the ophthalmic branch of the trigeminal accompanying the trochlear as it (ophthalmic) passes into the nasal region.

6. Lens. Remove the eyeball and cut off its dorsal side. The lens will be seen within as a whitish subspherical object just back of the pupil and iris. It is rounded toward the retina and is somewhat flattened toward the pupil.

7. The optic "coats." These are the outer, tougher *sclera,* the middle, darkly pigmented and vascular *choroid coat,* and the inner, light sensitive *retina* whose many nerve fibers converge at the *optic papilla* to form the optic nerve.

8. Optic humors. In the anterior and posterior chambers, on either side of the iris and between the cornea and the lens, is the *aqueous humor.* The jelly-like *vitreous* (glassy) *humor* fills the large chamber behind the lens.

C. THE EAR.

The turtle ear consists of two parts — middle and internal. The *middle ear* lies just medial to the circular *tympanic membrane* which is just caudal to the angle of the jaw. The *internal ear* lies medial to the middle ear and is surrounded by a bony capsule.

1. Columella, a slender straight bone attached at one end of the posterior-central region of the tympanic membrane and at the other end to the medial wall of the *middle ear cavity.*

2. Opening of auditory tube, a slit-like opening bounded by raised folds located ventral to the inner end of the columella. This leads into the *auditory* (Eustachian) *tube* which can be probed and found to open into the side of the mouth cavity.

The details of the inner ear are essentially the same as those of the shark's inner ear which should have been previously studied, therefore, the inner ear of the turtle will not be described in this manual. (Dissection of the inner ear is far more simple and satisfactory in the shark, Squalus sp.).